THE NORTH AMERICAN SWANS
THEIR BIOLOGY AND CONSERVATION

Fig. 1. Trumpeter swan, standing posture

The North American Swans

Their Biology and Conservation

Paul A. Johnsgard

School of Biological Sciences
University of Nebraska–Lincoln

Zea Books
Lincoln, Nebraska

2020

Abstract

Among birds, swans are relatively long-lived species and are also among the most strongly monogamous, having prolonged pair and family bonds that strongly influence their reproductive and general social behavior, which, in combination with their beauty and elegance, contribute to the overall high degree of worldwide human interest in them. This volume of more than 59,000 words describes the distributions, ecology, social behavior, and breeding biologies of the four species of swans that breed or have historically bred in North America, including the native trumpeter and tundra swans, the introduced mute swan, and the marginally occurring whooper swan. Also included are 5 distribution maps, 15 drawings, 27 photographs by the author, and a reference section of nearly 1,000 literature citations.

ISBN 978-1-60962-171-1

doi: 10.32873/unl.dc.zea.1100

Composed in Sitka types.

Zea Books are published by the University of Nebraska–Lincoln Libraries.

Electronic (pdf) edition available online at
https://digitalcommons.unl.edu/zeabook/

Print edition available from
http://www.lulu.com/spotlight/unllib

UNL does not discriminate based upon any protected status.
Please go to http://www.unl.edu/equity/notice-nondiscrimination

Find this and additional works at
https://digitalcommons.unl.edu/zeabook/
https://digitalcommons.unl.edu/biosciornithology/
Part of the Natural Resources and Conservation Commons, Ornithology Commons, Other Life Sciences Commons, Population Biology Commons, Poultry or Avian Science Commons, and the Terrestrial and Aquatic Ecology Commons

pajohnsgard@gmail.com

Nebraska
UNIVERSITY OF
Lincoln®

Dedicated to all those who revere and endeavor

to conserve the natural world,

in whatever ways they can.

Fig. 2. Whistling swan, adult in flight

Contents

Maps

Figures

Photographs

Preface

My first encounter with wild swans occurred in North Dakota during the mid-1940s, when I was about 14 years of age and was duck hunting with my father on a foggy October morning. Indeed, the fog hung so low it was impossible to see more than about 100 feet ahead or above, although I could easily hear the familiar calls of ducks and cattle all around me. Then I heard a new and strange sound that I thought must indicate approaching geese, and I nearly strained my eyes as I stared into the gray skies above. Gradually, out of the mists emerged about a dozen whistling swans, flying so low I could hear their wingbeats even before their ghostlike shapes emerged into view. I had never seen a wild swan, and I stood transfixed and speechless. Before I could think to alert my father, they were mystically out of sight, and as their voices faded in the distance, I was left behind with only an indelible lifetime memory.

Not long afterward I sold my shotgun to buy a camera, and my goals were redirected to learn about rather than to try to kill birds of such great beauty. I have since become ever more enchanted with waterfowl and have spent much of my life researching and writing about them. Over the past seven-plus decades, I have also been captivated by cranes, grouse, and various other groups of both North American and world birds. Although I have at times been thus diverted from studying and writing about only waterfowl, I have also gradually become emotionally reconnected to swans. I have camped on the bird-rich coastal tundra along the Bering Sea coastline of western Alaska to observe nesting whistling swans and other Arctic-nesting water birds, studied trumpeter swans breeding in their remote Rocky Mountain retreats of the Yellowstone–Grand Teton region, and rejoiced in the increasing numbers of these spectacular birds as they reclaim their ancestral homelands of central and eastern North America.

During the past half-century, the trumpeter swan has been rescued from possible extinction by massive private, state, and federal conservation efforts and has increased from some isolated populations totaling a few hundred individuals to numerous flocks comprising tens of thousands of birds. Because of federal restoration efforts that began regionally at South Dakota's Lacreek National Wildlife Refuge, a large breeding population of trumpeter swans has developed over the past five decades in the central Great Plains and the Great Lakes states. Thousands of trumpeter swans and some whistling swans now winter even in Nebraska, adjacent western Iowa, and along the Missouri River valley of northwestern Missouri. Seeing immaculate white swans flying above the autumn-tinted loess hills of the Missouri valley every fall, their artfully elongated body forms and wings outlined against cerulean skies, always emotionally transports me back to that wondrous day of epiphany in my North Dakota youth.

In 2016 I wrote *Swans: Their Biology and Natural History*, as part of a five-volume series of short digital books intended to update my earlier writings on North American ducks, geese, and swans, especially those species that have exhibited recent major abundance and distributional changes. This text builds on the base of the 2016 book but has been further updated and enlarged. In keeping with the geographic limits of my other four recent volumes on waterfowl, this work is restricted in coverage to the four swan species that are known to have bred in North America during historic times—the mute, trumpeter, tundra (whistling), and whooper—but has excluded three non–North American swan species.

In expanding and updating my previous coverage of the general biologies of these swans, I have given particular attention to their present population trends and related conservation issues. I have also included an extended bibliography of nearly 1,000 titles to provide readers insight into the abundance and variety of these important literature sources. Reading about swans isn't nearly so exhilarating or so emotionally powerful as seeing them in nature, but I hope that by reading this book people might gain a deeper awareness and understanding of these magnificent birds and make an effort to see them in their natural surroundings.

Paul A. Johnsgard
Lincoln, Nebraska

Foreword and Acknowledgments

This is the last of seven volumes on the biologies of North American waterfowl and gallinaceous birds that I have published through the University of Nebraska–Lincoln Libraries' Digital Commons and Zea Books imprint since 2016. The volumes include separate descriptive accounts of 55 species of wild waterfowl that have been documented as having bred or at least regularly occurred within the geographic limits of North America north of Mexico and within the Caribbean region, including 16 surface-feeding ducks, 15 sea ducks, 8 geese, 7 diving ducks, 4 swans, 3 whistling ducks, 2 perching ducks, and 2 stiff-tailed ducks. There are also 21 accounts of North American gallinaceous birds, including 12 grouse, 6 New World quails, 3 introduced partridges, and 1 introduced pheasant. Collectively the volumes include more than 526,000 words, 1,300 pages, nearly 300 behavioral sketches, 180 formal ink drawings, 155 photographs, and 76 range maps, all of my own making. Including duplicate citations that appear in two or more volumes, there are also more than 6,000 literature citations.

For all the work associated with the editing and assembling of these numerous elements into coherent books, words cannot express my enduring thanks to Dr. Paul Royster, Coordinator of Scholarly Communications for the University of Nebraska–Lincoln Libraries. His skills, and those of his steadfast and sharp-eyed editor, Linnea Fredrickson, have transformed my efforts, affected by increasingly unreliable memory and faltering eyesight, into objects of utility and beauty. I am forever grateful.

I am also deeply indebted to the School of Biological Sciences of the University of Nebraska–Lincoln for allowing me to retain my office facilities for two decades after retirement to continue my research and writing. Lastly, my thanks and apologies to my family, who have let me wander in search of birds to distant and sometimes dangerous parts of the world without uttering any complaints of my frequent absences during birthdays, graduations, and even the birth of a daughter.

Fig. 3. Whooper swan, adult wing-flapping

I. Introduction to the Northern Swans and Their Biology

Because of their immaculate white plumage and their strong pair and family bonds, swans have long served as icons of beauty, devotion, and longevity in the myths and folklore of many cultures. The four species of swans I have identified as North American swans include two abundant native species (trumpeter and tundra swans), one Eurasian species that was introduced and has become well established here during the past century (mute swan), and one relatively rare Eurasian visitor that has maintained a precarious toehold in the Aleutian Islands (whooper swan). In most ways all of these swans are very similar; one might say they are minor variations on an evolutionary theme first enunciated by Charles Darwin. Indeed, three of the taxa (trumpeter, tundra, and whooper) are so anatomically and biologically similar that their evolutionary relationships are still largely a matter of conjecture and likely to be resolved only by more detailed genetic research.

Most Americans are probably personally familiar with the regal-looking mute swan of Europe, which has long been associated with parks, zoos, and ponds on large private estates. When captive mute swans escaped from some Long Island estates during hurricanes in the late 1930s, many became feral. Their offspring plus those of other introductions have since expanded over much of the Atlantic Coast region, eventually occupying coastal wetlands from New Hampshire to the Carolinas. Beginning in the 1960s, introduced mute swans also spread out from estates along the eastern shoreline of Lake Michigan and have since occupied much of the Great Lakes region from Wisconsin to New York and southern Ontario. They are now classified as an invasive species in several Atlantic and Mississippi flyway states, in part because of the damage the birds cause to submerged aquatic vegetation beds, affecting the food base of other aquatic wildlife and disrupting the breeding sites of colonial nesting shorebirds. Also, overt attacks inflicted on people and pets when they venture too close to a nesting pair has sometimes has led to severe bodily harm of the intruder. Nevertheless, mute swans are undeniably beautiful and are likely to be the only species of swan familiar to most Americans, except those who live along major swan migratory routes or near important wintering areas.

The world's swans are part of the large and widespread waterfowl family of ducks, geese, and swans (Anatidae), which collectively total about 150 species. Most of the waterfowl are important game birds (about 15 million are shot annually in North America), although for aesthetic reasons the swans have traditionally been protected from hunting in nearly all advanced cultures. The United States is a sad exception to this generality; here the trumpeter swan by early in the twentieth century was shot to near-extinction for its flesh, feathers, and trophy values. Several thousand tundra swans still fall victim annually to federally sponsored "sport" hunting, as do some trumpeter swans, snow geese, and other waterfowl that some hunters seem to have difficulty distinguishing from tundra swans. Some of these same species die from lead poisoning years or decades later by ingesting lead shotgun pellets that have settled and accumulated on wetland bottoms.

All swans as well as geese, their closest relatives, are moderately to extremely large waterfowl. The mute swan is probably the heaviest of all flying vertebrates; zoo-raised males sometimes approach 50 pounds in body mass. Swans and geese are mostly associated with temperate to Arctic climates, where an abundance of terrestrial grasses and aquatic herbaceous plants provide a basic and usually unlimited food source. Swans and geese also all exhibit plumage patterns that lack brilliant coloration and are alike in both sexes, a reflection of long-term pair-bonding. Long-term bonding reduces the need for maintaining easily recognized clues to sex, age, and breeding condition that might be needed repeatedly over a lifetime but might also provide predators with information as to their location.

Swans differ from geese mostly in (1) being more dependent on submerged aquatic vegetation as a basic food resource, (2) being usually larger in body mass, and (3) having mostly or entirely white adult plumages (except for one of the world's seven species). In spite of their conspicuous plumages, most swans are also large and strong enough to protect themselves from all but the most formidable enemies, such as eagles and large canine predators.

The majority of the world's seven species of swans are found in the cooler parts of the northern hemisphere, the exceptions being three southern hemisphere temperate-zone swans. Most species of swans are seasonally migratory, exhibit prolonged periods of sexual immaturity (typically two to four years), and have strong monogamous pair-bonds that often persist for several years if not their entire adult lifetimes.

Swan family bonds are also strong and may likewise last until the offspring establish their own pair-bonds, or even longer. All swans are primarily vegetarians, obtaining most of their food from subsurface aquatic vegetation. The plumages of their downy young tend to be pale and simple, without strong patterning, and in most swans their adult plumages are also simple; all the North American swans are entirely white after molting their more grayish juvenile plumages during their first year of life. The immaculate white plumage of swans might be related to their effective visual advertisement signals associated with their usual high degree of territoriality.

All pair-bonded swans perform various pair-specific behaviors. Except perhaps for one aberrant species (the South American coscoroba swan, *Coscoroba*

coscoroba), these always include a mutual "triumph ceremony" (Fig. 7). This important pair-bonding display is most often performed after an aggressive encounter with another individual and has the outward appearance of representing a kind of victory celebration (Johnsgard, 1965).

All swans also perform generally highly stereotyped pre-copulatory and post-copulatory postural signals, the latter often accompanied by calling and wing-raising or wing-flapping. Such behaviors that perhaps help in strengthening pair-bonds tend to be much more similar among closely related groups than are the more highly species-specific pair-forming behaviors that might be more important in preventing interspecific pairing (Johnsgard, 1965).

The most highly territorial swans are also the most strongly vocal species. Three northern hemisphere swans—the trumpeter, whooper, and tundra—have the loudest and most penetrating voices. Their voices are amplified and resonated by virtue of the birds having extremely elongated windpipes (tracheae) that uniquely convolute within the keel of the breastbone (sternum). The resulting increased length and air-chamber volume provide enhanced resonating capacity and increased harmonic development for vocalized specificity and diversity (Fig. 4).

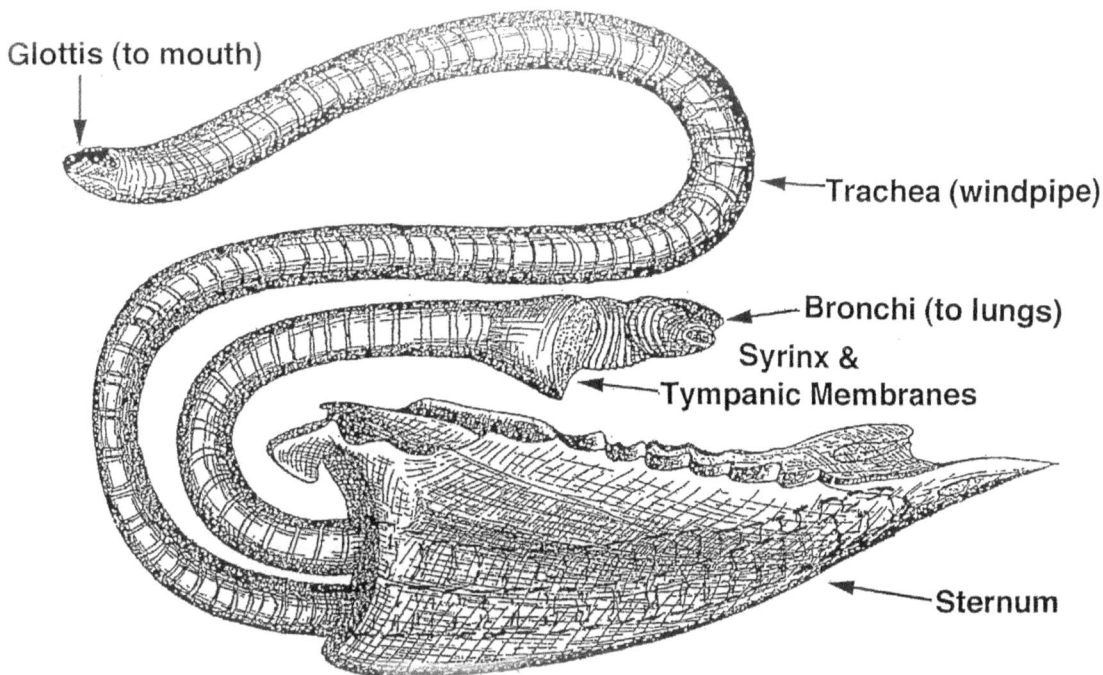

Fig. 4. Tracheal and external sternal anatomy of adult Bewick's swan

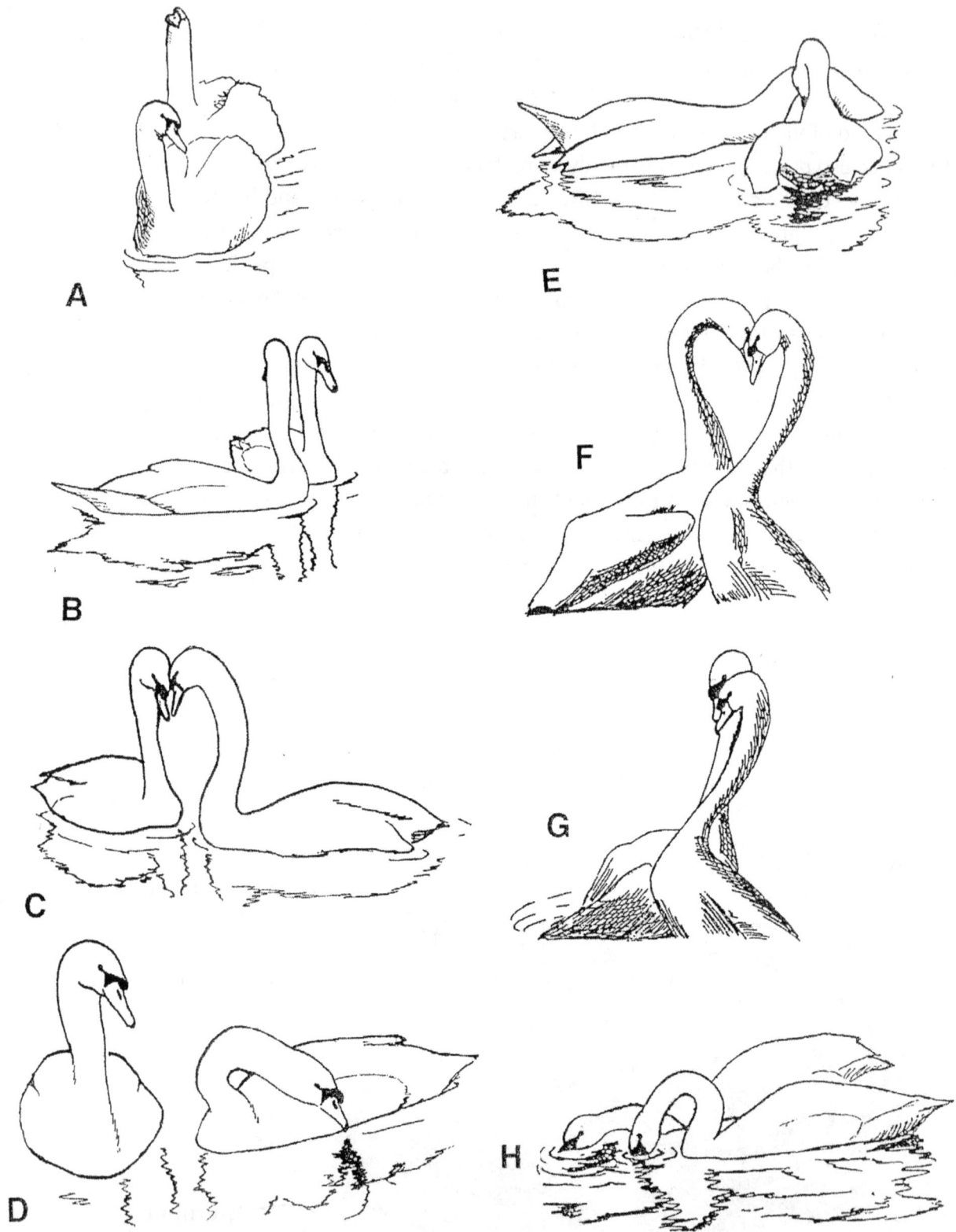

Fig. 5. Behavior of the mute (A–G) and (H) trumpeter swans. A. Adults in threat posture, rear bird chin-lifting. B–C. Pair head-turning. D–E. Precopulatory head-dipping. F–G. Post-copulatory display. H. Pre-copulatory head-dipping. (After Johnsgard, 1965.)

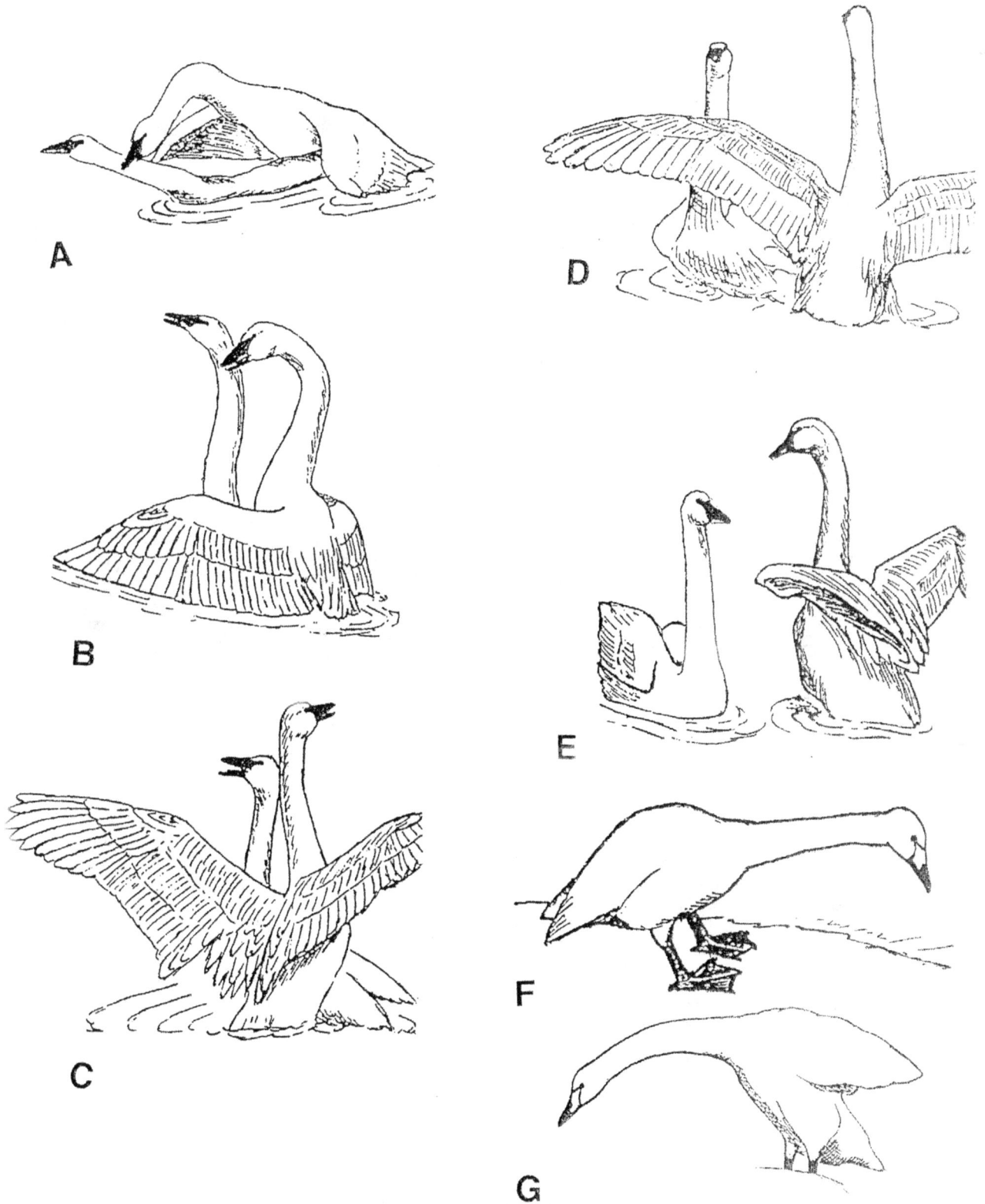

Fig. 6. Behavior of the trumpeter (A–E) and whooper (F–G) swans. A–E. Copulation sequence. F. Male threatening in bowsprit posture while calling. G. Male performing general shake during threat posture. (After Johnsgard, 1965.)

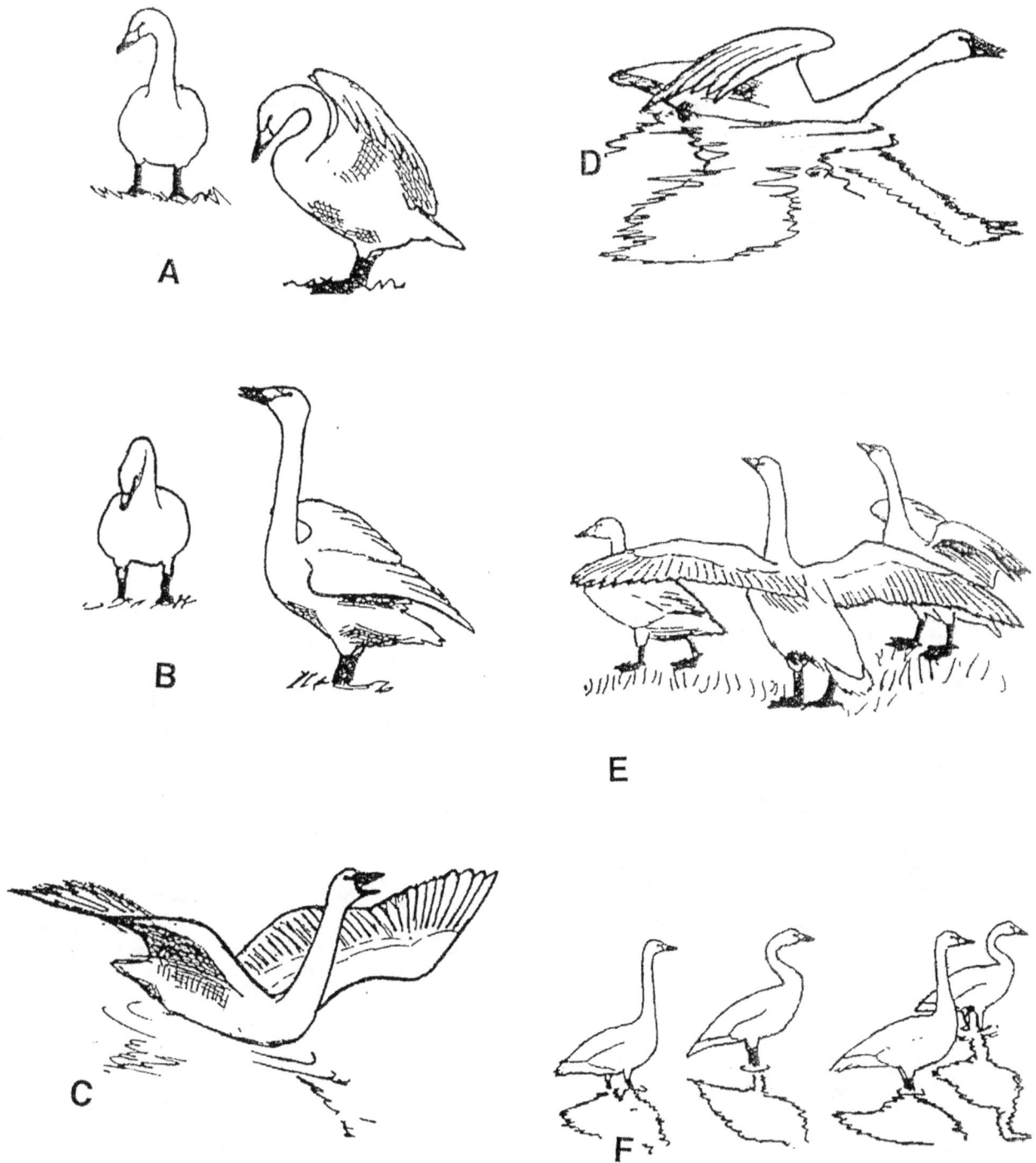

Fig. 7. Behavior of the whooper (A–B), whistling (C–D), and Bewick's (E–F) swans. A–B. Triumph ceremony, male at right. C–D. Male threat-calling while wing-waving (C) and while holding wings outstretched (D). E. Triumph ceremony by pair and a juvenile. F. Preflight neck-bending by family group. (After Johnsgard, 1965.)

In swans as well as in geese, the voice-producing structures are a pair of highly vibratory ("tympanic") membranes at the junction of the trachea and the two bronchi that connect directly with the lungs (Johnsgard, 1972). Through varied muscular tensions on these paired membranes, complex sounds are generated by expelled air, much like sounds produced by double-reed musical instruments. Swan vocalizations are very similar in both sexes, differing mainly in harmonic overtone characteristics and to a lesser degree in overall pitch (a mixture of generated sound frequencies). There is little variation in pitch among swan vocalizations, but they are often rich in acoustic structure (fundamental frequencies and harmonic overtones), and their utterances also often vary greatly in loudness, note cadence, and duration.

Adults of nearly all swan species produce a wide variety of vocalizations, including defensive hisses, parental contact notes, and preflight and in-flight calls as well as greeting and triumph ceremony calls between pairs (Cramp and Simmons et al., 1977; Limpert and Earnst, 1994; Mitchell and Eichholz, 2019). In most swan species such pair-specific vocalizations probably provide enough acoustic specificity in the form of varied sound frequencies, harmonic content, and such temporal differences as call duration and repetition rate as to allow for interindividual recognition among mated pairs and families and perhaps even larger social groups.

At the other extreme, the mute swan has highly limited vocal abilities, which in adults are mostly confined to soft hisses, grunts, and groans, and they lack the long-distance carrying power of the other northern swans. Its breeding territories are correspondingly relatively small, and in some protected locations mute swans even form rather dense nesting colonies ("swaneries") of up to about 100 birds.

The biological meanings and significance of most waterfowl vocalizations and other social behaviors are still largely mysteries, as are many other aspects of swan biology, such as within-family recognition abilities and the behavioral mechanisms for the development of flocking coordination and long-distance migratory traditions. Such are among the attractions of many species of birds, but in swans they are encapsulated within the bodies of some of the most beautiful, most graceful, and most entrancing of all birds on earth.

Fig. 8. Mute swan, adult male in flight

II. Species Accounts

Mute Swan *Cygnus olor* (Gmelin) 1789

Other vernacular names

White swan, Polish swan, Höckerschwan (German), cygne muet (French), cisne mudo (Spanish)

Subspecies

No subspecies recognized. A variant plumage called the "Polish swan" is a mutation-based color morph having a white rather than grayish downy and juvenile plumages, and pinkish legs and feet that persist from the downy stage into adulthood.

Range

Introduced and local but gradually expanding west into locales in the Midwestern states (Michigan, Wisconsin, eastern Iowa, northeastern Missouri, Illinois, Indiana, northwestern Ohio, western Pennsylvania) and south along the Atlantic coast from New Hampshire to Virginia, and locally to South Carolina. Some seasonal movements extend southward to Kentucky and Tennessee. Pioneering birds have occurred west to Saskatchewan, Minnesota, South Dakota, and Nebraska. Breeds locally in Europe as feral or semiferal flocks in Great Britain, France, Holland, and central Europe. Breeds under native wild conditions in southern Sweden, Denmark, northern Germany, Poland, and Russia; also southward in Turkey and Iran east through Afghanistan to Mongolia. In winter occurs south to northern Africa, the Black Sea, northwestern India, and Korea. Also introduced and locally resident in South Africa, Australia, and New Zealand.

Measurements and weights

Wing: Males 589–622 mm; females 540–96 mm. *Culmen:* Males 76–85 mm; females 74–80 mm (Scott and the Wildfowl Trust, 1972). *Eggs:* Eggs average 115 × 75 mm, greenish blue, 340 g. Bauer and Glutz (1968) summarized available data.

    Weights: Males 8.4–15.0 kg (avg. 12.2 kg); females 6.6–12.0 kg (avg. 8.9 kg). Males seldom weigh more than 13.5 kg (29.7 lb) and females not much over 10 kg (22 lb). However, four old birds weighed between September and December averaged 16.22 kg (35.8 lb) with a maximum of 22.4 kg (49.4 lb). Scott and

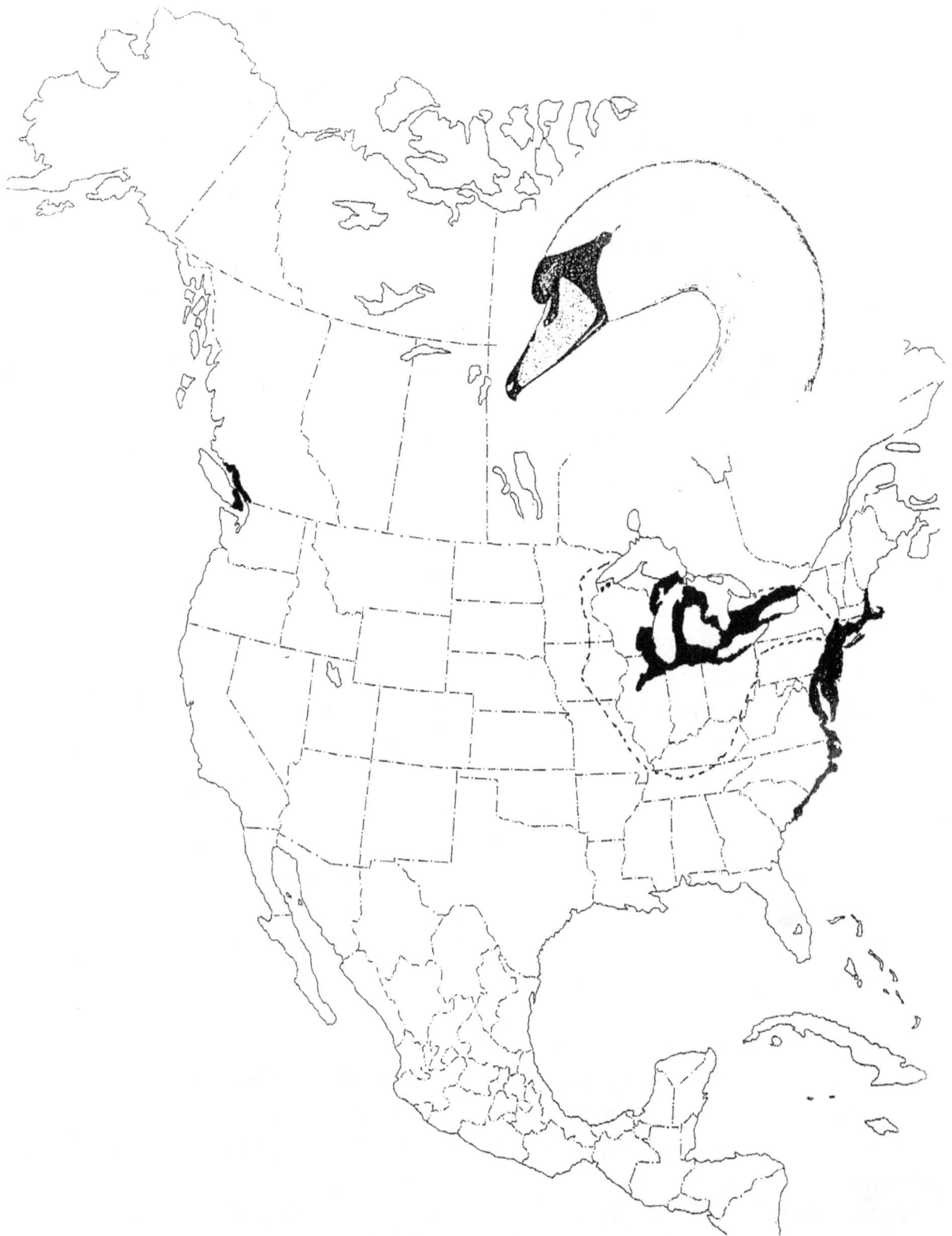

Map 1. Introduced North American breeding or residential (inked) distribution and wintering limits (dashed line) of the mute swan as of 2019.

the Wildfowl Trust (1972) presented weight data indicating that although male mute swans average slightly heavier than male trumpeters (12.2 kg vs. 11.9 kg [26.8 lb vs. 26.2 lb]), female mute swans average slightly lighter than female trumpeters (8.9 kg vs. 9.4 kg [19.6 lb vs. 20.7 lb]).

Identification and field marks

Length 50 to 61 inches (125–155 cm). Adults are entirely white in all post-juvenile plumages. The bill is orange with black around the nostrils, nail, and edges of the mandible. The feet are black, except in the uncommon "Polish" color phase, in which they are fleshy gray. Females are smaller (see Measurements and weights) and have a less fully developed knob over the bill. Juveniles exhibit a variable number of brownish feathers, which diminish with age (except in the Polish swan variant, which has a white juvenile plumage), and the fleshy knob over the bill remains small through the second year of life. Mute swans are the only white swans that have generally reddish to orange bills adorned with an enlarged black knob at the base (lacking in immatures), outer primaries that are pointed toward their tips, and a long, somewhat pointed tail shape. The trachea, unlike those of native North American swans, does not enter the sternum.

Males are considerably heavier and larger than females, and individuals in excess of 10 kilograms are most probably males. Males also have larger black knobs at the base of the bill and more often assume the familiar threatening posture. For immature birds, internal examination is required to determine sex. Any bird still possessing feathered lores (the area between the eyes and bill) or some brownish feathers of the juvenal plumage is less than a year old. Second-year birds may have smaller knobs and less brilliant bill coloration than is typical of older birds.

In the field: This large swan is usually seen in city parks but may occasionally be found as a feral bird under natural conditions, especially in the eastern states and provinces. The neck of the mute swan seems thicker than those of the trumpeter and whistling swans, and while swimming the bird holds it gracefully curved more often than straight. Further, the wings and scapulars are raised when the birds are disturbed, rather than being compressed against the body. The orange bill and its black knob are visible at some distance. In flight, the wings produce a loud "singing" noise that is much more evident than in the native North American swans, and, additionally, mute swans rarely if ever call when in flight, as is so characteristic of the native species. A snorting threat is sometimes uttered by male mute swans, which is their apparent vocal limit.

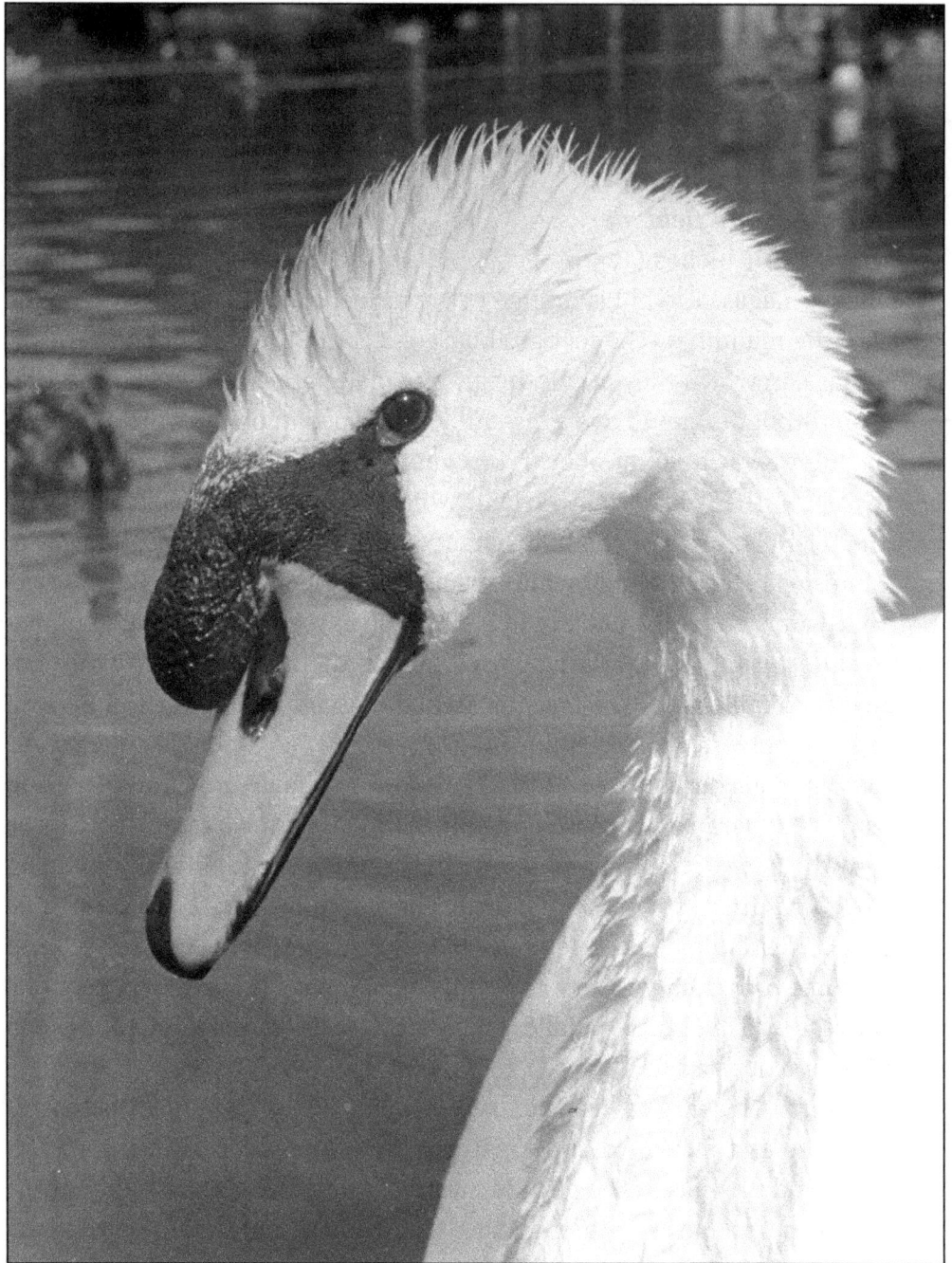

Mute swan, male displaying mild threat

Evolutionary Relationships

In spite of its similar geographic range and plumage characteristics, the mute swan is not closely related to the other northern hemisphere white swans, but instead its nearest relative is the Australian black swan (*Cygnus atratus*), as is indicated by a variety of behavioral characteristics (Johnsgard, 1965). It has hybridized in captivity with trumpeter, whooper, and tundra swans (Johnsgard, 1960; Owens, 1995; Brazil, 2003), in the wild or semiwild conditions with the whooper swan (Brazil, 2003), and apparently in the wild with the trumpeter swan (Trumpeter Swan Society, n.d.)

Natural History

Habitat and foods

The habitats of this species are diverse and numerous, reflecting a long association with humans and at least partial domestication. Probably, however, temperate zone marshes, slowly flowing rivers, and lake edges were its native habitats. Clean, weed-filled streams are preferred over larger polluted rivers. Brackish or saltwater habitats are often used during nonbreeding periods.

A variety of studies of foods taken in Europe and North America indicate that the leafy parts of many freshwater or saltwater plants, such as pondweeds, muskgrass, eelgrass (*Zostera*), and green algae (*Enteromorpha*), are major sources of food where they are available. Also where it is available, waste grain is often consumed, and some grazing on terrestrial grasses near water has become increasingly prevalent in many areas (Scott and the Wildfowl Trust, 1972). Mute swans can reach underwater foods up to several feet (up to 50 cm) below the surface by upending, but in common with other swans they do not dive when foraging. A small amount of animal foods, including amphibians, worms, mollusks, and insects, may be taken when available, but these are a minor part of their diet.

Willey (1968a) estimated that adults eat an average of 8.4 pounds of vegetation per day. In general, the birds feed on subsurface plants they can reach when swimming or by tipping up in the manner of dabbling ducks. In England these foods include algae (*Chara, Enteromorpha, Ulva, Nitella*), pondweeds (*Zostera, Potamogeton, Ruppia*), grasses, and other herbaceous plants. Some terrestrial vegetation is also consumed, and sometimes small aquatic animals, including fish and amphibians, were reported in the diet.

Population History

Except in a few areas where local foraging conditions favor large concentrations, as on some coastal islands, mute swans are not notably social. Nevertheless, they may flock in groups of a thousand or more in summer molting areas, and similar flocking may also occur in the winter. In many parts of its range the mute swan is essentially sedentary, and in England, for example, studies of banded birds have revealed that most movements are less than 30 miles and usually follow watercourses (Ogilvie, 1967).

Mute swans, adult pair

In North America mute swans occupy a sedentary breeding and wintering range with habitats that are directly related to human presence. There seems to be no historical account of the spread of the species in the Hudson River valley and on Long Island after it was originally released as a park bird. Being properly considered an exotic, the species was not included in official bird lists until the 1930s, when the fourth (1931) edition of the American Ornithologists' Union's *Check-list of North Americans Birds* first noted that it had become established in the lower Hudson valley and on the south shore of Long Island, sometimes straying to the coast of New Jersey.

East Coast hurricanes, such as one that occurred in 1939, caused additional dispersal of birds previously confined to wealthy estates on Long Island and in Rhode Island. By 1949 the species had spread throughout much of Long Island

and had also become well established in Rhode Island. By the late 1950s it was nesting along the entire shore of Rhode Island, and breeding had been reported in the District of Columbia.

A secondary population center was then independently developing on upper Lake Michigan around Grand Traverse Bay and Lake Charlevoix (Edwards, 1966). Early counts of this population were reported by Banko (1960), who noted an increase from two birds in 1948 or 1949 to 41 by 1956. Apparently initiated by a release of two birds in 1918, the flock consisted of at least 600 by 1973. Attempts have been made to establish new flocks in Illinois, Ohio, Arkansas, Oklahoma, Texas, and New Mexico. The species has also expanded into southeastern Ontario along Lake Erie and Lake Ontario; Kingston had the largest number of mute swans (737) reported from any count circle in Canada during the 2017–18 Audubon Christmas Bird Count.

In 1962 mute swans were introduced into the Chesapeake Bay region when five birds escaped from an estate in Maryland. By 2002 they had increased to about 4,500, at which time the entire Atlantic flyway population was estimated at about 12,600 birds (Perry, Osenton, and Lohnes, 2004). From 1986 to 2002, the population in the Atlantic flyway more than doubled from approximately 6,300 birds to more than 14,000. In response to concern about the swans' deleterious ecological effects on wetlands, the Atlantic Flyway Council adopted a Mute Swan Management Plan in 2003, and many states implemented management programs that reduced the flyway-wide population to about 9,000 birds by 2011 (Costanzo et al., 2015).

There is also an isolated population at the southern end of Vancouver Island and in the Fraser River valley of British Columbia that was initiated in the late 1800s and had become well established by the 1950s.

In recent years, pioneering birds have occupied new localities for breeding. These include many nestings in the New England states plus isolated breeding records in Nebraska, South Dakota, Saskatchewan, and elsewhere in southeastern Canada and the eastern United States.

Social Behavior

Flocking occurs among nonbreeders and unsuccessful breeders during the mid-summer molting period, and later in the fall these flocks are increased by the addition of family groups forced out of their territories by cold weather. Atkinson-Willes (1963) indicated eleven locations (mostly coastal) where aggregations of more than 250 swans have regularly been reported in Great Britain. Except where local foraging conditions favor such large concentrations, as on some coastal islands, mute swans are not notably social, nor are they highly mobile.

Probably the most extensive migrations occur among the breeding birds of Siberia and Mongolia, from Lake Baikal east. It is presumably those birds that winter along the Pacific coast, from Korea south to the Yangtze Kiang, suggesting a migration of up to about 1,000 miles. Pairs and family units migrate together and remain together until about the end of the year, when at least in England the breeding adults begin to exhibit territorially. At that time the young birds remain in the winter flocks of nonbreeding birds, and they may remain together through the following summer and winter (Scott and the Wildfowl Trust, 1972).

Minton (1968) has studied population densities in England and reported a density of one pair (about 30 percent nonbreeders) per 5.5 square miles on his study area of 550 square miles. He noted that this represented about one breeding pair per eight square miles, compared with earlier estimates of one pair per 16 square miles reported for England and Wales as a whole. The highest reported county densities were one pair per three square miles for Middlesex and one per seven square miles in Dorset. Atkinson-Willes (1963) reported that the famous mute swan colony at Abbotsbury in Dorset averaged 66 pairs of breeding swans (range 39–104) during the years 1947 to 1956 and had an average total population of about 700 birds. A tradition of protection and abundant food in the form of *Zostera* and *Ruppia* accounted for this concentration of birds.

Comparable breeding density figures are not available for North America, but the highest recent US Christmas counts have occurred in Rockwood, Michigan, with as many as 1,141 birds seen in a 15-mile-diameter area (176 square miles), or 6.2 birds per square mile. Canadian locations have tallied as many as 1,201 birds at Holiday Beach, Ontario, or 6.8 birds per square mile. Willey (1968a) estimated that 24.5 to 54.3 percent of the Rhode Island population represented potential breeders. If Minton's similar estimate that 30 to 40 percent of the population consists of breeding birds, this would represent a breeding density of nearly two pairs per square mile, assuming no spring dispersal. It would further seem that, at least locally, mute swan breeding populations in North America may be as high as or higher than those in Great Britain.

In Europe the mute swan is a species that nests largely in populated areas that support few other breeding waterfowl, and there is probably little competition with other species. Although Dementiev and Gladkov (1967) considered it to be tolerant toward other birds and sometimes occurring with nesting graylag geese (*Anser anser*), in contrast, Stone and Masters (1971) reported that six captive mute swans killed six adult geese and two adult ducks as well as 40 ducklings and goslings during a 20-month period. Willey (1968a) stated that nesting birds sometimes kill other swans that intrude into their nesting areas and considered mute swans to be a substantial threat to humans, particularly children. There are cases of mute swans causing the death of a child and two adult humans by drowning (Baldassarre, 2014).

Mute swans are highly sedentary birds in Great Britain. Atkinson-Willes (1963) reported that only a small number of banded mute swans had been proven to move more than a hundred miles, and only two had been known to cross the English Channel. In many parts of its range the mute swan is essentially sedentary, and in England, for example, studies of banded birds have revealed that most movements are less than 30 miles and usually follow watercourses (Ogilvie, 1967). Later, Harrison and Ogilvie (1968) noted that ten of 2,700 band recoveries exhibited overseas movement from Great Britain, with recoveries from the Netherlands, Sweden, France, and the Baltic coast of Germany. Many of these recoveries were related to severe winter conditions that forced birds to move between the continent and Britain.

According to Minton (1968), most movements of mute swans occur before their mating and acquisition of a territory, after which they become quite sedentary. Most pairs return to their territory year after year, with only 2 percent of the surviving paired population that Minton studied moving their territories more than five miles. Nonbreeding pairs and unsuccessful breeders frequently move to the nearest flock for molting in midsummer, while unsuccessful breeders molt on their territories and move into flocks during fall. Among paired birds, movements are usually less than ten miles, and only about 5 percent of the 450 pairs Minton studied moved farther than this. However, unsuccessful breeders are more likely to move greater distances than successful ones.

Flocking typically occurs among nonbreeders and unsuccessful breeders during the midsummer molting period, and later in the fall these flocks are increased by the addition of family groups forced out of their territories by cold weather. Unusually large flocks have been seen on a 1,240-acre reservoir at Abberton, a summer molting area that attracts up to nearly 500 birds maximally, and along the Essex coast at Mistley, where 800 to 900 birds were attracted to waste corn from a mill.

Reproductive Biology

Pair-forming behavior occurs in the fall and winter, usually during the season prior to a bird's initial breeding. Among mute swans, initial breeding is most frequent in the third year, with some birds (mostly females) breeding when two years old and some not until four or older. Pair formation and pair-bonding in this species, as in all typical swans, occurs by mutual greeting ceremonies such as head-turning, and bonds are firmly established by triumph ceremonies between members of a pair. This display occurs after the male has threatened or attacked an "enemy" and returned to his mate or prospective mate with ruffled neck feathers and raised wings, calling while chin-lifting.

Copulatory displays have been described by various persons, such as Boase (1959), Johnsgard (1965), and others. Precopulatory displays involve mutual bill-dipping and preening movements, with the neck feathers ruffled. Following treading, both birds rise in the water breast to breast, with their necks and heads extended vertically but their wings closed; then they gradually arch their necks and settle back on the water (Fig. 3). Precopulatory behavior may occur frequently among paired birds in winter flocks.

Once formed, pair-bonds are very strong, and so long as both members of a pair remain alive there is a low rate of mate changing. One study in England indicated that there was a 9 percent "divorce" rate among unsuccessful breeders or nonbreeders, and a rate of less than 3 percent among successful breeders (Minton, 1968). Minton found that the maximum known duration of a pair-bond was 17 years in a male and 16 years in a female. The maximum known longevity for wild birds is 26 years and 9 months (Baldassarre, 2014).

The earliest known age of reproductive maturity in North America has been reported as two (Johnston, 1935) or three (Willey, 1968a) years, but studies in England indicate considerable variation may occur. Minton (1968) reported on the initial pairing behavior of 125 mute swans of known age. Nearly half of these were two-year-olds, another 30 percent were three-year-olds, and a few (one male, four females) took mates when only a year old. Most birds were paired for at least a year before they actually attempted to nest, with only 2 of 60 birds that were no more than two years old actually nesting that year. Birds tended to pair with others of about their own age, with a slight tendency for the males to be older than the females. Further, in 74 percent of the initial pairings neither partner had ever been paired before. Birds pairing for the first time with a previously paired bird were generally replacements for dead mates.

Perrins and Reynolds (1967) indicated that three years of age is the most common time of initial breeding for females, but a few birds may breed at two and some may not breed until they are six years old. Initial breeding by males occurred between three to seven years of age. Minton (1968) found that of 43 mute swans, half initially nested and raised young at the age of three, while an additional third did so the following year, with a slight tendency for females to mature earlier than males. Three birds did not breed until they were at least six years old.

The strong pair-bond of all swans is well known and well documented in mute swans. Minton (1968) reported that "divorce" (the changing of partners when both birds are still alive) among the paired population was low.

Mute swan, male displaying swimming threat

In cases where both birds survived to following years, 82 percent of the successful breeders and 78 percent of unsuccessful breeders remained paired. Of 71 pairings first studied in 1961, six were still intact in 1966. During the six-year study, 11 males and 9 females were known to have had at least 3 different mates, but in several cases (12 males and 2 females) birds that had apparently lost their mates remained on their nesting territory the following year. In some cases there was a gap of two or three years before re-pairing, while in others the birds apparently gave up pairing permanently.

The nesting period of mute swans occurs in spring, which is generally March through June in its northern hemisphere range, and from September through January in New Zealand and Australia. Mute swans are territorial, or at least are highly defensive of their nest sites, although the size of the territory varies inversely with breeding density. In England, nests are most often placed in or near standing water, less often beside running water, and least often in coastal situations. Nests are sometimes grouped in colonies—one colony in Dorset had as many as 500 nests (Scott and Boyd, 1957). One study in Staffordshire, England, indicated a density of a pair per 2,000 hectares (4,900 acres) but with defended territories limited to a few meters of streamside locations. In a few areas of dense colonies, the nests may be located only a few meters apart.

Established breeders tend to use previous nest sites. Willey (1968a) estimated the average size of 12 nesting territories as 4.4 acres (range 0.5–11.8) in Rhode Island. Minton (1968) noted that both breeding and nonbreeding pairs were more prevalent on small (ten acres or less) water areas than on larger ones, but considering availability, larger water areas were slightly favored. Likewise, streams were favored over canals or rivers more than 20 feet wide, especially by breeding pairs. Clean waters with aquatic vegetation were also preferred over more polluted waters.

Most studies indicate that six eggs constitute a modal clutch size for mute swans. Studies summarized by Bauer and Glutz von Blotzheim (1968) indicate averages of between 5.8 and 6.2 eggs. Clutches of up to 11 eggs laid by one female have been reported, but renesting attempts appear to average about 4 eggs (Perrins and Reynolds, 1967).

After the establishment of a breeding territory, nests are constructed on land or shallow water. The nests are usually about a meter in diameter and 0.6 to 0.8 meter in height and are constructed in the form of a large mound of vegetation consisting of rushes, reeds, other herbaceous vegetation, and sometimes also sticks. The nest cup is lined with finer materials and also with down and feathers. The female typically does most of the nest construction, but the male also gathers material from nearby, passing it back toward the nest over his shoulder. Down-plucking may begin with the start of egg-laying, the initiation of incubation, or not until the last or penultimate egg is deposited.

The female does the incubation but is closely guarded by the male. The cygnets typically leave the nest on the day after hatching and remain closely attended by both parents, often riding on the back of one or both parents. Similar back-riding has been reported only once in other North American swans (Bailey, Bangs, and Bailey, 1980). The wing molt of both parents normally occurs during the fledging period of the brood (Bauer and Glutz von Blotzheim, 1968; Dementiev and Gladkov, 1967). Nests are large piles of herbaceous vegetation, built by both sexes, and often built on the previous year's nest, especially if it has been a successful site. Eggs are deposited every other day until a clutch of four to eight (usually five to six) eggs is laid. Incubation is normally performed only by the female, but the male may occasionally take over for a time.

The incubation period has generally been estimated as 35 or 36 days, with some estimates of up to 38 days (Bauer and Glutz von Blotzheim, 1968). The female incubates, but the male actively protects the nest. Minton (1968) reported a 59 percent nesting success among 352 pairs, and a 52 percent success rate for 11 renesting attempts, with 80 percent of the nest losses due to human disturbance or destruction. Willey and Halla (1972) reported the loss of 87 eggs and young from a total of 47 nests after severe flooding and cold weather in Rhode Island.

The fledging period has been variously reported as four and a half months (Bauer and Glutz von Blotzheim, 1968), 18 weeks (Lack, 1968b), 18 to 20 weeks (Scott and Boyd, 1957), and 120 to 150 days (J. Kear, in Scott and the Wildfowl Trust, 1972), during which time the adults undergo their postnuptial molt. Successful breeders remain with their young well past the cygnets' fledging time, usually until severe weather forces the families to retire to winter quarters, where they associate with larger groups of swans.

Typically, the young of the past year are driven out of the territory by their parents before the latter begin to breed again. Minton (1968) reported two cases in which young remained with their parents until the following summer or until molting, and in neither case did the parents breed during that year.

Minton observed two cases of pairing between parents and offspring. One involved the pairing of a female with her yearling son after the male parent had died, while the other involved a female observed paired with a two-and-a-half-year-old son. In neither case did actual nesting occur.

Minton (1968) found that the average brood size (219 broods) at fledging over a six-year period was 3.5 birds, while the total number raised to fledging averaged 2.0 per breeding pair. Perrins and Reynolds (1967) found an average brood size of 3.1 young for 83 broods, with an estimated 2.0 young raised per pair to September, including those pairs that failed to hatch any young. These authors estimated that the average mortality rate between hatching and fledging was 50 percent, with an additional 23 percent mortality rate for the rest of the year. Willey (1968a) estimated a similar prefledging mortality of 56.4 percent in 1968, with the snapping turtle (*Chelydra serpentina*) apparently a primary predator of cygnets.

After fledging, the family increases their food intake and fat reserves before leaving the breeding grounds. Perrins and Reynolds (1967) estimated that among immature birds there is a 67 to 75 percent survival (25 to 33 percent mortality) rate, whereas breeding adults have a survival rate of 82 percent, which possibly decreases after the sixth year of life. There is little difference in the estimated mortality rates of the two sexes. Ogilvie (1967) estimated a higher mortality rate of 40.5 percent for birds banded when under a year old and 38.5 percent for those banded when over a year old, with possibly greater survival chances in the third and fourth years of life than during the first two. These survival rates are similar to more recent estimates provided by McCleery et al. (2002).

As with other species of swans, in-flight collisions with overhead wires or other objects have been found to be a major cause of mortality, with oiling, disease, fighting, cold weather, and shooting also accounting for some deaths. The maximum longevity record for a banded, free-living mute swan is apparently 19 years. The longest documented survival record in captivity is 21 years (Scott and the Wildfowl Trust, 1972), although captives have reputedly lived for as long as 40 years.

Status

The mute swan has greatly benefited from human influence and has either been purposefully introduced or otherwise spread into new breeding areas of North America since it was introduced in the late 1800s. In many areas of Europe, where the species was exterminated during the nineteenth century, it has now become reestablished and is increasing, with an approximate recent population of 250,000 birds (Rees et al., 2019).

The annual Breeding Bird Survey provides estimates of population trends for many species of North American birds. The long-term data (1966–2015) for the mute swan indicates a survey-wide annual population increase of 1.79 percent, but with astonishing rates of increase of 17.13 percent in Canada, Ontario, and the lower Great Lakes and St. Lawrence region, 15.57 percent in Pennsylvania, and 11.08 percent in Massachusetts (Pardieck et al., 2018).

The annual Christmas Bird Counts of the National Audubon Society also provide a rough index to the population growth rate of mute swans in North America since the late 1940s. During the years 1949 through 1969, the numbers of such counts approximately doubled from 403 to 876, while the total number of mute swans counted more than quintupled from 374 to 1,644. The average total count for the ten-year period 1950–59 was 504 birds, with an average of fewer than 20 stations reporting the species, whereas during the period 1960–69 the average total North American count was 1,434 birds, with an average of 34 stations reporting mute swans. The long-term (1966–2017) national population trends in Audubon Christmas counts as of 2017 (https://www.audubon.org/conservation/where-have-all-birds-gone) have been annual increases of 4.24 percent in the United States and 11.18 percent in Canada (Meehan et al., 2018). During the 2017–18 Christmas count, a single count circle in Michigan (Rockwood) reported 1,141 mute swans present, almost as many as was the average number for the species counted *nationally* during the 1960s.

Mute swan, male in flying attack

Midsummer surveys have been conducted in the Atlantic flyway at three-year intervals since 1986, when 6,383 swans were tallied. A maximum total of 14,344 swans were seen in 2002, which had declined (as a presumed result of control efforts) to about 9,000 in 2011 (Cornely, Petrie, and Hindman, 2014). These Atlantic flyway swan counts were summarized for each state by Costanzo et al. (2015). Estimated state total swan populations in 1986 and 2011 were as follows: Ontario ("Lower Great Lakes"), 615, 3,062; Maine, 3, 0; New Hampshire, 19, 7; Vermont, 0, 0; Massachusetts, 585, 1,046; Rhode Island, 880, 778; Connecticut, 1,452, 809; Delaware, 21, 41; New York, 1,815, 1,765; Pennsylvania, 137, 167; Maryland, 264, 76; Virginia, 60, 241; West Virginia, 0, 21; North Carolina, 3, 30; South Carolina, 3, 30; Georgia, 0, 0; Florida, —, 100. Comparable data for the Mississippi flyway are lacking.

Surveys suggest that the entire North American mute swan population was in the vicinity of 22,000 to 25,000 birds early in the twenty-first century (Baldassarre, 2014). As of 2013, Michigan had the largest reported statewide population with about 17,000 swans. Because the birds have caused environmental damage—such as damage to native aquatic vegetation and undesirable effects on native water birds—Michigan agencies have developed plans for reducing the state's population to 9,000 birds by 2030, although the state's estimated population was still about 15,000 in 2019. A recent collective North American population estimate was 50,000 to 60,000 mute swans, and the species' overall native Eurasian population was about 642,000 birds (Rees et al., 2019).

Trumpeter Swan *Cygnus buccinator* (Richardson) 1758

Other vernacular names

Trumpeter, wild swan, Trompeterschwan (German), cygne tromette (French), cisne trompetero (Spanish).

Subspecies

No subspecies are recognized. Native breeding populations currently exist in southern and central Alaska from the Gulf of Alaska north to the Brooks Range, southwestern Yukon and Northwest Territories, northern British Columbia, western and southern Alberta, and western Manitoba. In the western United States, native populations breed in eastern Idaho, southwestern Montana, and Wyoming. Reintroduced and breeding locally at national wildlife refuges in south-central Oregon (Malheur), eastern Washington (Turnbull), and central Nevada (Ruby Lake). Also, reintroduced populations breed eastward locally in the Great Plains from eastern Manitoba, South Dakota, and Nebraska eastward through Minnesota, Ontario, Iowa, Wisconsin, Ohio, Michigan, New York, and Pennsylvania. Colonizing birds might occur farther east in the New England states. Some movements occur in winter south to California, New Mexico, Texas, and the Gulf Coast.

Range

Drewien and Benning (1997) stated that wintering trumpeter swans have been visually reported from Chihuahua and Tamaulipas, Mexico, and that there have been band recoveries from Chihuahua and Durango. Overseas vagrants have been reported from as far away as England (Dalgleish, 1880), Japan (Murase, 1993), and Russia (Syroechkovski, 2002).

Weights and measurements

Wing: Adult male 545–680 mm (avg. 618.6 mm); adult female 604–636 mm (avg. 623.3 mm). *Culmen:* Adult male 104–119.5 mm (avg. 112.5 mm); adult female 101.5–112.5 mm (avg. 107 mm) (Banko, 1960). *Eggs:* Average 118 × 76 mm, white, 325 g.

Weights: Hansen et al. (1971) presented weight data on ten wild adult males that averaged 11.97 kg (26.4 lb) with a range of 9.5 to 13.6 kg (20.9–30.0 lb), and 11 adult females that averaged 9.63 kg (20.6 lb), with a range of 9.1 to 10.4 kg (20.1–2.9 lb). Scott and the Wildfowl Trust (1972) reported the average weight of ten captive males as 11.9 kg (26.2 lb), with a range of 9.1–12.5 kg (20.1–27.5 lb). Banko (1960) reported that the minimum weight of eight

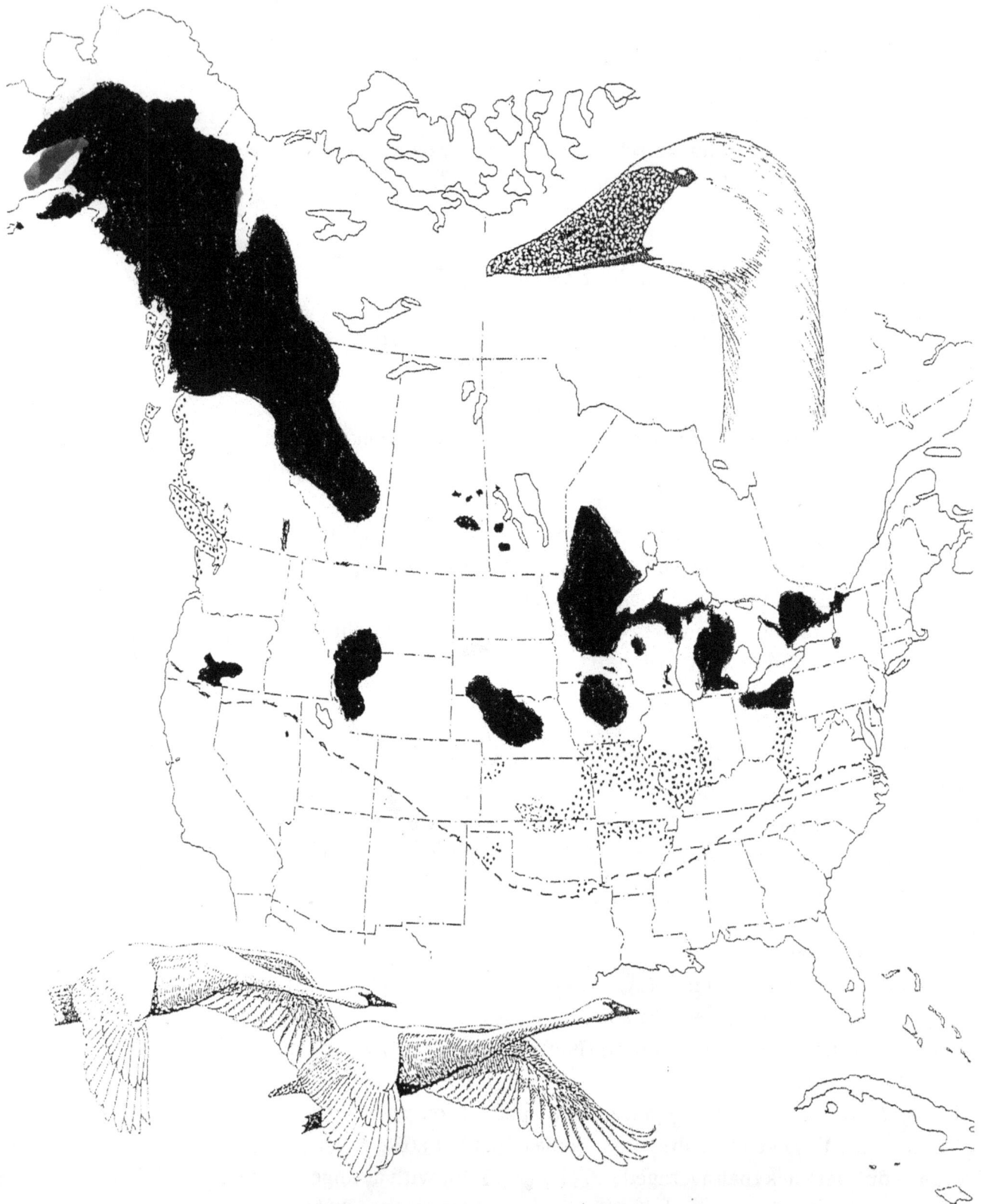

Map 2. Native and reintroduced breeding ranges (inked) of the trumpeter swan, as of 2019. Major wintering areas are stippled, and a dashed line indicates the approximate southern limits of wintering.

wild males at least two years old was 20 lb (9.07 kg), whereas the minimum weight of 14 females of similar age was 16 lb (7.25 kg). Eight males at least one year old had a minimum weight of 18 lb (8.16 kg), and four females of this age had a minimum weight of 15 lb (6.8 kg). Nelson and Martin (1953) indicated an average weight of seven (presumably wild) males as 27.9 lb (12.65 kg), with a maximum of 38 lb (172 kg); the average of four females was 22.5 lb (10.25 kg), with a maximum of 24.5 lb (11.16 kg). Eleven females averaged 9.4 kg (20.7 lb), with a range of 7.3 to 10.2 kg (16.1–22.5 lb) (Drewien and Bouffard, 1994).

Identification and field marks

Length 60 to 72 inches (150–180 cm). This swan's huge size, its entirely white plumage, and the absence of any yellow on the bill should serve to identify it. If the bird weighs more than 20 pounds, measures at least 50 mm from the tip of the bill to the anterior end of the nostril, and has entirely black lores or at most a pale yellow or gray mark on the lores, it is most probably a trumpeter swan. The bill of adults usually measures at least 50 millimeters from the front edge of the nostril to the tip of the nail. The length of the middle toe (excluding claw) is at least 135 mm, compared to 133 mm maximum in whistling swans (Banko, 1960). Drewien and Bouffard (1994) compared partial bill length (bill tip to nostril), middle-toe length, and tarsal length measurements of trumpeter and tundra swans and found that among adults bill measurements (bill tip to anterior edge of nostril) provided the lowest percentage overlap (1–20 percent), between the two species, followed by middle toe length (17–42 percent), and tarsal length (92–100 percent).

As noted in the whistling swan account, the dorsal surface of the sternum should be examined to be absolutely certain of species identification; the presence of a dorsal protrusion near the sternum's anterior end is the best criterion of a trumpeter swan. The total length of the trachea from glottis to the bronchi averaged (in two cases of personally observed dried specimens) more than 130 centimeters. Hinds and Calder (1973) reported a tracheal length of 121 to 135 centimeters and a tracheal volume in trumpeter swans that was 3.6 to 4.0 times greater than would be expected in a nonconvoluted trachea. In contrast, the sternum of an adult whistling swan, although penetrated by the trachea, lacks a mid-dorsal protrusion, and the trachea measured (in one dried specimen) under 120 centimeters (Fig. 9).

Females are identical to males, averaging only slightly smaller in most measurements. Juveniles exhibit some gray feathers in their plumage up to about a year from hatching; thereafter immature birds are essentially identical to adults in appearance. The grayish plumage of the juvenile is held during most of its first year of life, and the lores are likewise feathered for much of the first year.

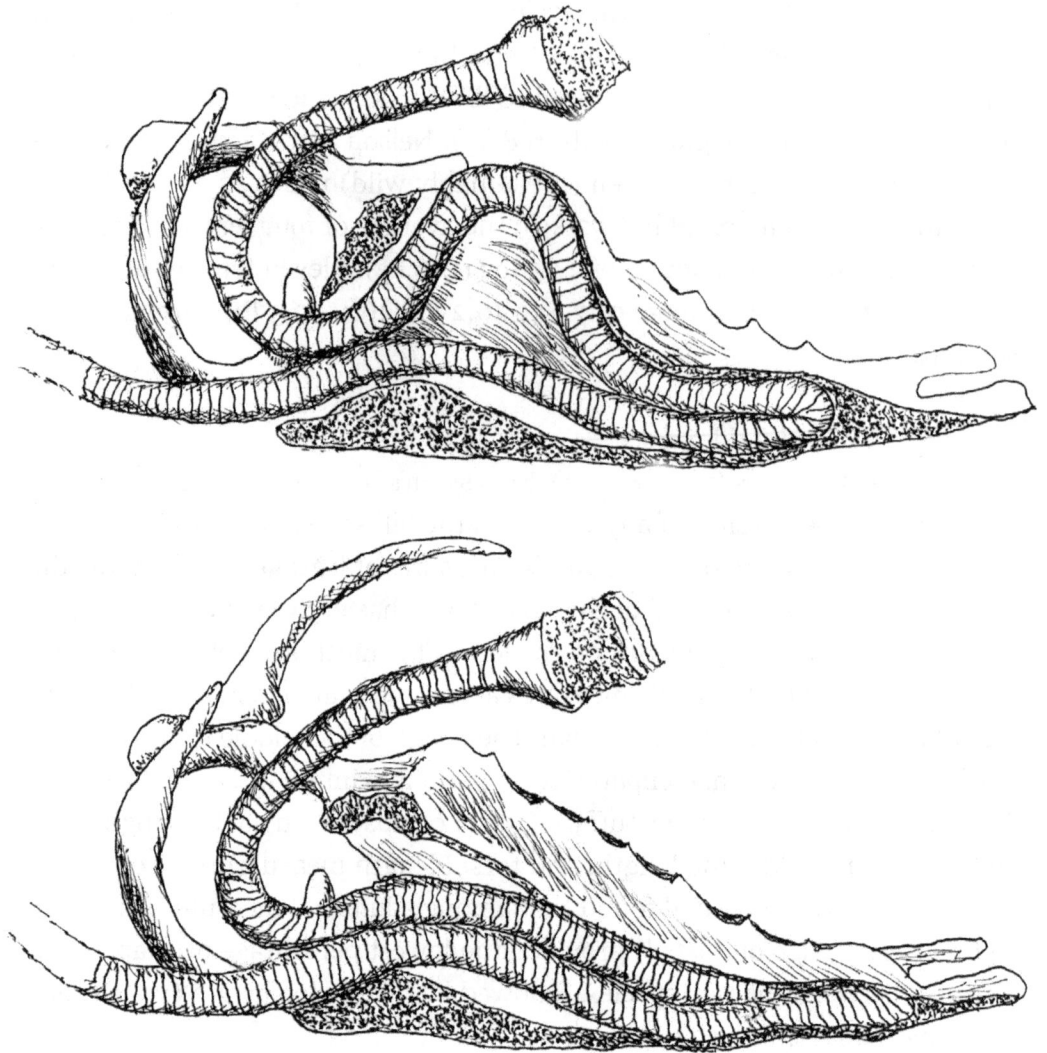

Fig. 9. Sternal anatomy of trumpeter swan (above) and whistling swan (below). (After Kortright, 1943.)

Although the birds are usually pure white at the age of 12 to 13 months, a few dark feathers may persist somewhat longer (Hansen et al., 1971). Second-year immature birds may perhaps be distinguished from older birds on the basis of their incompletely developed sexual structures. Also, the forehead feathers of immature birds extend forward to a point on the culmen, while these feathers in adults have a more rounded anterior border.

In the field: In the field, the absence of definite yellow coloration on the lores and a voice that is sonorous, hornlike, and dual-note—often sounding like *ko-hoh*, which can be heard for a mile or more under favorable conditions, rather than being higher pitched and a series of single notes, sounding like a barking *wow, wow-wow*—are the most reliable field marks for trumpeter swans. Wood,

Brooks, and Sladen (2002) observed that the single-syllable calls of trumpeter swans are lower pitched (fundamental frequency of 303 Hz) than those of the tundra swan or trumpeter × tundra hybrid swans. The closely spaced harmonics visible on a sound spectrogram might number as many as 22 on some loud calls (Peiplow, 2019). Hansen et al. (1971) stated that a more nearly straight culmen (the profile of the dorsal edge of the upper mandible) typical of this species (especially adult males), as compared with a more concave culmen in the tundra swan, might also provide a useful clue for field identification.

Numerous records of whooper swans (*Cygnus cygnus*) have been made in mainland North America, especially along both coasts. Thus, a remote possibility exists for encountering this species, which is slightly smaller than the trumpeter swan and has a bright yellow area of bare skin that extends from in front of the eyes (the lores) out onto the bill, where it reaches the nostrils and extends slightly beyond (see whooper swan species account for more details).

Evolutionary Relationships

The trumpeter, whooper (*Cygnus cygnus*), whistling (*C. c. columbianus*), and Bewick's (*C. c. bewickii*) swans collectively constitute a close-knit evolutionary complex referred to as the "northern swans," in which the species limits and internal relationships are still far from clear (Johnsgard, 1974). González, Düttmann, and Wink (2009) reported that, based on genetic data, *Cygnus cygnus* and *C. columbianus* had a more recent time of genetic divergence than the separation time of *C. olor* and *C. atratus*), and their separation probably occurred during Pleistocene times (Sun et al., 2017). Ploeger (1968) similarly concluded that the present distributions and geographic variations among extant Arctic ducks, geese, and swans should be ascribed to the physical-geographic situation (the presence of available refugia) during the Last Glacial Period.

Given the conspecificity of the whistling and Bewick's swans, it seems intuitive that the trumpeter swan and whooper swan are each other's nearest relatives (Johnsgard, 1974), although Livezey (1996) judged the Bewick's swan, rather than the whooper swan, to be the nearest relative of the trumpeter swan. Some breeding sympatry occurs between the trumpeter and whistling swans in north-central Alaska, where, with a warming climate, trumpeter swans have moved into areas previously occupied only by whistling swans (Wilk, 1993; Bryant, Scotton, and Hans, 2005; Loranger and Lons, 1990; Harwood, 2010), in the same way that in Asia whooper swans have been moving north to occupy areas that had previously been occupied by Bewick's swans (Brazil, 2003). No definite evidence of hybridization between the trumpeter and whistling swan in the wild has appeared, but captive-bred hybrids between them have been documented (Wood, Brooks, and Sladen, 2002), and they are known to be fertile

Trumpeter swans, adult pair

(Mitchell and Eichholz, 2019). No proven case of hybridization yet exists between whooper and Bewick's swans in the wild, although captive hybrids have been reported (Rees, 2006).

In captivity several other hybrid combinations of the northern swans have been reported: tundra swans have been hybridized with mute swans, and whooper swans have hybridized with mute, trumpeter, and tundra swans (Johnsgard, 1960; Brazil, 2003). More surprisingly, two juvenile trumpeter × mute swan hybrids have been photographed with their presumptive parents near Vancouver, British Columbia, and were described as a "wild mating" (Trumpeter Swan Society, n.d.).

Natural History

Foods and habitats

Although small cygnets rely on high-protein animal foods such as aquatic insects and crustaceans, they progressively shift to a vegetable diet as they grow older. Banko (1960) summarized data on trumpeter swan foods in Montana and reported use of foliage and tubers of pondweeds (*Potomogeton*), water milfoil (*Myriophyllum*) leaves and stems, pond lily (*Nuphar*) seeds and leaves, water buttercup (*Ranunculus*) leaves, and a variety of additional herbaceous aquatic and semi-aquatic foods, such as *Chara, Anacharis, Lemna, Scirpus, Sparganium, Carex,* and *Sagittaria.*

When feeding in shallow waters, trumpeters use their strong legs and large feet to excavate the tubers and rhizomes of various aquatic plants, often forming large holes on the bottoms of marshes, such as at the Red Rock Lakes National Wildlife Refuge (NWR) in Montana. They also swim with the head and neck under water, pulling rooted materials off the bottom of the ponds. Additionally, they are readily able to remove duckweed (*Lemna*) or other small foods from the water surface by straining them through the bill in the manner of dabbling ducks, and they may feed heavily on duckweed when it is available.

Parental "puddling," a characteristic rapid paddling of the feet during swimming, apparently serves to stir up food from pond bottoms (De Vos, 1964). This behavior was observed by De Vos to be mostly performed by an adult female, occasionally by its mate, and several times by a cygnet. Female swans of various species frequently perform this behavior when leading broods, apparently thus improving the foraging efficiency of the short-necked and weak-legged cygnets. I have also observed surface-feeding ducks, such as mallards, being attracted to this behavior, foraging as close to the swan as possible.

Banko (1960) considered the presence of permanently open water with associated aquatic vegetation, a certain amount of level and open terrain, and a minimum of heavy timber near watercourses as important features of winter trumpeter swan habitat. The breeding habitat found at Red Rock Lakes NWR was characterized by Banko as consisting of large shallow marshes or shallow lakes of high fertility and at moderate elevations (ca. 6,600 feet elevation), with a profusion of aquatic plants of submerged and emergent growth forms, and with generally nontimbered but well-vegetated shorelines. In nearby Yellowstone National Park, the breeding lakes are mostly deeper, more heavily timbered, and higher in elevation (Yellowstone Lake is ca. 7,700 feet elevation) and therefore the location represents more marginal breeding habitat. Conditions are somewhat better in Grand Teton National Park, where the elevations of major breeding wetlands are lower (ca. 6,300 feet at Flat Creek Marshes near Jackson), and some of the marshes more closely resemble those of Red Rocks Lakes NWR.

Banko (1960) characterized the breeding habitat of trumpeter swans as stable waters that lack marked seasonal fluctuations; quiet waters of lakes, marshes, or sloughs that are not subject to current or constant wave effects; and shallow waters that allow for foraging on aquatic plants. Originally, the birds bred over a large area of North America's northern or western regions, including boreal forests, montane pine forests, the western edge of the eastern deciduous forest, the short-grass plains, and tall-grass prairies. Obviously the surrounding habitats are less important than the water characteristics, which must provide security, reasonable seclusion, and adequate foraging and nesting opportunities.

The species' primary foods consist of the leaves and stems of such aquatic plants as pondweeds and crowfoot (*Ranunculus*) as well as the roots of pondweeds and tubers of arrowhead (*Sagittaria*). Emergent shoreline plants are also consumed, and in late fall the seeds of sedges (*Carex*) and water lilies (*Nuphar*) become important foods. During winter the tubers of *Potamogeton* pondweeds are important foods, as are the leaves and stems of waterweed (*Elodea*) and muskgrass (*Chara*). Cygnets forage to a large extent on animal life in their early weeks, especially on aquatic insects and small mollusks, especially snails (*Physella*).

In some areas, such as western Washington, pastures are foraged, and crop plants such as corn, potatoes, carrots, and winter wheat are consumed, especially during winter, while newly planted crops, such as corn and soybeans, have been reported as spring foods.

Recent Populations (1950s to 2015)

Banko's 1960 report on the status of the trumpeter swan indicated that fewer than 200 breeding birds were then known to occur in the entire United States. Many of the known birds were located at Red Rock Lakes NWR, in the high-altitude and remote Centennial Valley of southwestern Montana, which was established in 1932. The lowest elevation at the refuge is about 6,600 feet, where several large marshes and cold-water lakes occur.

During the years 1954 to 1957 in Red Rock Lakes NWR, an average of 13 nesting pairs occupied Upper Lake (2,880 acres), 51 pairs occupied River Marsh (8,000 acres), and 15.5 pairs occupied Swan Lake (400 acres), representing

Trumpeter swans, pair with downy cygnets

a total average breeding population of about 80 pairs on 11,280 acres, or 4.5 pairs per square mile (140 acres per pair).

Besides the Red Rock Lakes–Yellowstone–Grand Teton population, other major nesting populations were later found in western Canada and Alaska. For example, Marshall (1968) reported that a nesting population at Grande Prairie, northwestern Alberta, numbered about 100 birds. Mackay (1957) judged that a small population of trumpeter swans that were then known to breed in the Peace River district of northern Alberta migrated to the northern United States and mixed with swans from the Red Rock Lakes NWR during winter months.

By the 1960s the Comox Valley of eastern Vancouver Island was also known to have a population of wintering trumpeter swans, but the breeding areas of those and other swans wintering in western British Columbia were then unknown. Hansen et al. (1971) confirmed that the wintering birds of western British Columbia represented part of a previously unrecognized Alaskan breeding population. There, the birds were found to be nesting commonly along the southern coast from Yakutat east to Cordova and in the Copper River drainage. Additional Alaskan breeding grounds were later discovered in the Kantishna, Tanana, Yukon, Susitna, and Koyukuk river valleys and on the Kenai Peninsula (King and Conant, 1981; Conant et al., 1986, 1988, 1992, 1994, 1999, 2002, 2007).

Hansen et al. (1971) provided an early numerical estimate of this species' North American population, reporting that they found a total of nearly 3,000 swans during aerial surveys in 1968. There was then thought to be about 150 additional swans that summered in western Canada and more than 300 in the vicinity of Red Rock Lakes NWR (representing a breeding population of about 80 pairs). Some swans also occurred in Yellowstone and Grand Teton National Parks, of which 59 birds were concentrated in Yellowstone. There were then about 600 additional birds present in other US refuges and zoos in 1970, suggesting that more than 4,000 wild and captive trumpeter swans then existed in North America (Denson, 1970; Hansen et al., 1971).

After later extensive surveys of their breeding grounds, the US Fish and Wildlife Service and Canadian Wildlife Service eventually recognized three major North American breeding populations of trumpeter swans, which they named the Pacific Coast Population, the Rocky Mountain Population, and the Interior Population. The Interior Population of the Great Plains and eastern North America had been wholly extirpated by the early 1900s as a result of habitat loss and uncontrolled hunting, but these regions may have once supported as many as 100,000 swans.

The Pacific Coast Population extends from the Cook Inlet and interior wooded river valleys of central Alaska east to the Pacific drainages of Yukon Territory and northern British Columbia. Its historic breeding range is still essentially intact. The Rocky Mountain Population extends from northern Yukon Territory and adjacent Northwest Territories south through northern British Columbia and northern Alberta to about the US-Canada boundary. Prior to the start of reintroduction programs in the 1950s, the population at Red Rocks Lakes NWR and the Greater Yellowstone region was the only remaining breeding population south of Canada.

The US component of the Rocky Mountain Population began to recover in the 1950s through local transplants of unfledged birds from Red Rock Lakes NWR. This program has generated several new breeding sites in the western United States. Thus, swans were introduced in Malheur NWR in central Oregon during 1955, with the first successful breeding during 1958 and increased numbers until the 1980s, when a decline began. Other reintroduction efforts at Sumner Lake, Oregon; Lower Klamath and Modoc refuges, California; and Turnbull NWR, eastern Washington achieved marginal success but have not flourished.

After 1954 releases at Ruby Lake NWR, northeastern Nevada, breeding success there first occurred during 1958, and the locale attained a peak population of 36 birds in 1975 before declining. Later introductions were made in the Flathead and Blackfoot valleys of western Montana. Although the Washington, Oregon, and Nevada introductions have not been successful, the two Montana introductions had built up to a population of 170 adult and subadult birds by 2015. However, the historic native population in the Greater Yellowstone region has undergone some declines in recent years, and the swan's future as a breeding bird in Yellowstone National Park is questionable (Smith and Chambers, 2011; Shields, 2019).

The US breeding segment of this flock is currently centered in the tristate region of eastern Idaho, southwestern Montana, and the Greater Yellowstone region of western Wyoming. This tristate flock had an estimated 530 adult and subadult birds in 1968. After a slow decline the tristate component of the Rocky Mountain Population had about 550 adult and subadult birds in 2015. Other components resulting from restoration efforts were then present in Montana (180 birds), Oregon (24), Washington (8), California (2), and Nevada (2).

The major segment of the Rocky Mountain Population of trumpeter swans consists of the large and flourishing Canadian portion that breeds from the eastern Yukon and adjacent Northwest Territories southeast to eastern British Columbia and central Alberta. In 1968 its population was estimated at less than 100 birds, but by 2015 had reached nearly 11,000 adult and subadult swans. About 3,000 have been wintering during recent years at Red Rock Lakes NWR, many of the thousands from the Canadian segment of the Rocky Mountain Population.

The Interior Population, which once bred across the Great Plains and eastward to the Great Lakes, was totally extirpated during settlement times, but has recently been recreated by transplant programs. It has since been largely restored as a result of successful reintroduction efforts in several states and provinces. Beginning in 1960, young birds from Red Rock Lakes NWR were released in Lacreek NWR in South Dakota. The first successful nesting from this effort occurred in 1963 (Monnie, 1966; Marshall, 1984). Offspring from Lacreek NWR gradually self-expanded into the Nebraska Sandhills (Ducey, 1999). Other birds now breed from the High Plains to Ontario and New York. By 2015 the western (High Plains) component of this still-expanding population was centered mostly in Nebraska and South Dakota and numbered nearly 500 birds.

Introductions were also begun in the Minnesota's Hennepin County Park District in 1966 as well as elsewhere in the state beginning in 1982, and these efforts have been the locus for the establishment of a major Minnesota breeding population. These birds were later supplemented by releases in northern Iowa, starting in 1995. Michigan began a similar release program in 1986, followed by Wisconsin in 1987. Other introductions in southwestern Ontario from 1983 to 2006 have also been successful (Lumsden et al., 1994; Lumsden and Drever, 2002). Nesting by the introduced birds has been reported as far south as Missouri (Neptune, 2007).

Trumpeter swans have increased remarkably in North America since the 1970s. As an amazing example of conservation success, the currently expanding Pacific Coast Population, which breeds from Alaska to northwestern British Columbia and the Northwest Territories and winters from southern Alaska to Washington and Oregon, numbered 24,240 birds in 2015 (Groves, 2017).

More than 27,000 trumpeter swans made up the eastern component (Mississippi and Atlantic flyways) of the still rapidly expanding Interior Population by 2015, with subpopulations in Minnesota (17,021 birds), Wisconsin (4,695), Michigan (3,021), Ontario (1,471), Iowa (204), Ohio (154), and New York (22). In recent years a few trumpeter swans have been venturing south and east to their historic wintering areas on the Gulf Coast and Chesapeake Bay. By 2015 the total North American trumpeter swan population was estimated at 63,016 adult and subadult birds (Groves, 2017). Adding in the unknown number of juveniles, the actual population might have reached 100,000 birds.

General Behavior

Only during the winter season are trumpeter swans appreciably social, and then the limited areas of open water force a degree of sociality upon them. The majority of flocks that are seen during the winter at Loess Bluffs NWR, northwestern Missouri, are groups of up to six birds, nearly all of which appear to be in a pair or family. Banko (1960) noted that it is seldom that more than six or eight swans fly together during local flights unless they are simultaneously

Fig. 10. Trumpeter swan, adult in flight

flushed. He included a photo of 80 birds occupying a small spring in mid-January but mentioned that as early as February pairs and small flocks begin to spread out over the snowfields that overlie their breeding habitat.

As noted earlier, the average Red Rocks Lakes NWR breeding density between 1954 and 1957 was 4.5 pairs per square mile (142 acres per pair) in three major nesting habitats, and in the most favorable nesting habitats about 70 acres per nesting pair was recorded during one year (Banko, 1960). Banko indicated that birds occupying open shoreline usually defended more area than did those nesting on islands, but shoreline nesters sometimes defended only a small bay area around the nesting site. Hansen et al. (1971) suggested that spatial isolation rather than food supply or size of area was important in determining territorial boundaries.

Trumpeter swans have no significant contact with tundra swans on their breeding or wintering areas, and Banko (1960) reported that they are highly tolerant of other bird and large mammal species. Even among pairs on their breeding territory, the presence of geese, pelicans, cranes, or herons is usually not sufficient to cause aggression, although swans leading young are less tolerant than others. One case was found, for example, of a nesting swan's killing a muskrat (*Ondatra zibethicus*) that approached a brood.

De Vos (1964) described the daily activity patterns of three captive swans, which may not be wholly typical of wild birds. He noted that the pair performed bathing, preening, sleeping, loafing, swimming, and foraging several times daily and usually in unison. Preening bouts typically followed bathing and lasted for varying periods up to 85 minutes. In total, the adult pair slept about the same amount of time during the egg-laying period, while later in the summer a month-old cygnet slept more than the total of both parents. In general, preening most commonly occurred early in the morning, early in the afternoon, and during the evening. Feeding occurred after the morning and evening preening periods, reaching a maximum in early afternoon, with a secondary evening peak.

There are few good data on daily movements, but Monnie (1966) reported that local movements of up to about 100 miles were noted at Lacreek NWR over a prolonged period. Banko (1960) observed that flights during local movements were usually at lower altitudes than were longer flights. At Loess Bluffs NWR, the flocks that are moving about on the 3,500-acre marsh rarely fly as high as 100 yards. During the past few years, it has been common to see flocks of 150 to 200 birds foraging or resting on an area of open water at Loess Bluffs in midwinter when the total refuge population might exceed 1,000 birds, but the groups are usually scattered widely over the marsh, especially where the water is about three feet deep and aquatic plants such as lotus (*Nelumbo lutea*) are abundant.

Reproductive Biology

Social behavior

Large flocks of trumpeter swans are now becoming increasingly common as the national population increases but suitable habitat declines. Banko (1960) provided a photograph of a flock of 80 birds at Montana's Red Rock Lakes NWR on a small spring in January, at a time when the refuge population numbered less than 200 birds. Such groups consist of families and pairs, the family bonds probably persisting through at least two years, judging from a limited amount of information on banded birds. A good deal of aggressive behavior and mutual display occurs in wintering flocks, and much of it is concerned with pair formation and pair-bond maintenance.

Pair-bonds are probably usually formed in the second winter of life, or possibly the third winter in some individuals. Banko (1960) reported that the greatest incidence of social display occurs in fall when new birds join others on the wintering grounds and in spring just before dispersal to the breeding grounds. Triumph ceremonies, with attendant wing-waving and calling by both birds as they partially rise in the water, are the most common of these displays; copulatory behavior is more likely to occur on the nesting territory just before egg-laying. Precopulatory behavior (Figs. 5 and 6) involves mutual head-dipping, which closely resembles bathing movements, by both members of a pair, and treading is followed by loud calling, wing-waving, and rising in the water in a manner closely resembling the triumph ceremony (Johnsgard, 1965).

Dispersal to the breeding grounds from the wintering areas may involve movements of only a few miles, as on the western refuges, or a migration of several hundred miles, in the case of the breeding population of southern Alaska. On the Copper River and the Kenai Peninsula, spring arrival is in late March or April (Hansen et al., 1971). Because of their long fledging period, the birds must begin nesting at the earliest possible moment; thus, on their Alaskan range they are among the earliest breeding birds to arrive and the last to leave, establishing territories and nest sites on muskrat houses while these locations are still largely snow covered and iced in (Wilmore, 1974).

Pair-forming behavior

Banko (1960) summarized evidence that nesting may begin as early as the fourth year of life or as late as the sixth year, but it would seem probable that these examples are atypical and that initial nesting in the third year of life would be characteristic of wild birds.

At least in some cases, the birds may form pairs when 20 months old and begin nesting as early as 33 months after hatching (Monnie, 1966). In Alaska most breeding occurred when the birds were four to seven years of age, although a few pairs bred when two or three years old (Hansen, 1973). A wild-caught pair in the Philadelphia Zoo first nested successfully in 1965, although the female (of unknown age) had been in the zoo since 1959.

Monnie (1966) reported that courtship among 20-month-old swans began in mid-January and continued until mid-March, during which time among nine birds two apparent pairs were formed, plus a trio involving two males and a female, while two females remained unpaired. Monnie did not specifically indicate whether this courtship consisted of actual copulatory behavior or was pair-bonding behavior in the form of prolonged individual associations and mutual triumph ceremonies. Like other swans, trumpeters are monogamous and have strong pair-bonds, although some variations have been observed among captive birds. Banko (1960) reported a single case of a trio living together, although the sex of the extra bird was not learned. Griswold (1965) also reported a captive trio, in which a male was paired with two females.

Banko (1960) described the triumph ceremonies of this species, which are often performed following the expulsion of a territorial intruder. Triumph ceremonies in swans consist of an excited calling, wing-waving, or shallow wing-flapping with extended neck as the pair stands close to one another. They are most often performed after the expulsion by the male or both pair members of a real or perhaps imagined threat. Banko (1960) noted that such mutual displays regularly occur in the wintering areas among birds in flocks, although he did not clearly associate this behavior with pair formation. Triumph ceremonies involving more than two birds are common and probably represent participation by the past season's offspring, based on my observations at the Wildfowl Trust in England.

De Vos (1964) observed 11 copulations in captive trumpeter swans, 10 of which were seen between April 16 and 26 (the first egg was laid April 21). One copulation was also seen in mid-July, more than a month after hatching had occurred. As with other swans, it begins with increasingly coordinated bathing (head-dipping) movements by the two pair members, followed by the male mounting the female and clutching her outstretched neck. Typically, as copulation ends both sexes rise together in the water, variably extending their wings (the male usually more fully extending his) while both birds call in unison. Finally, the birds flap their wings once or twice, followed by bathing and preening (Johnsgard, 1965).

Nesting

At Red Rock Lakes NWR, muskrat houses are the most commonly chosen nest site, and territorial limits are likely to encompass large areas with many suitable nest sites. A nesting density of about 4.5 birds per square mile, or

Trumpeter swans, pair with downy cygnets

142 acres per pair, is average, with favorable areas supporting up to one pair per 70 acres. Birds nesting on islands defend smaller areas than those nesting on shorelines, and a feeling of spatial isolation rather than food supply is probably the major determinant of territorial size. Two-year-old pairs might establish territories, even though actual nesting is not attempted. Territories at Red Rock Lakes averaged 70 to 150 acres (28–61 hectares) (Banko, 1960). In Alaska, wetlands ranging in size from 5 to 128 acres (2.1–52 hectares) support only a single pair of swans (Hansen et al., 1971). Territories may be occupied as early as February, although eggs are not laid until the end of April or early May. Egg-laying there is initiated in late April or early May, and incubation is begun before the end of May (Hansen et al., 1971).

Nest sites and territories are often retained over several breeding seasons. At Red Rock Lakes NWR 82 of 199 nest sites were reused the following year (Banko, 1960), and in Yellowstone National Park a territory was used for an average of 14.7 years, with a maximum of 38 years (Proffitt et al., 2010), the latter figure suggesting that it was inherited by successive generations.

Nests are built by both sexes, usually on prior nest sites, and clutches of four to eight but most commonly five eggs are laid. Eggs are laid on alternate days, and the nest is strongly guarded by both sexes. Normally, only the female performs incubation, but a few instances of apparent incubation by males have been observed in captivity in this species.

The normal incubation period for trumpeter swans is 33 to 37 days, based on observations in Alaska and Montana and records from captivity. During this time the eggs may be left unattended for varying periods, but their large size probably reduces the rate of heat loss and possible chilling. However, even when the female does not cover the nest, the male closely watches it for possible danger, and little if any egg predation occurs under natural conditions. The cygnets feed on a diet rich in invertebrates and grow rapidly, especially during the long days on the Alaskan breeding grounds.

Banko (1960) reported on 109 nests observed over four seasons at Red Rock Lakes NWR. More than 70 percent of these were located on or very near a previous nest site, with four sites being used all four years. Island sites were preferred over shorelines, and fairly straight shorelines tended to be avoided. The highest concentrations occurred where irregular shorelines combined with numerous sedge islands to produce maximum habitat interspersion, resulting in maximum nest densities of one nest per 70 acres.

Hansen et al. (1971) noted that 32 of 35 Alaskan nests were built in water 12 to 36 inches deep, and 21 of 40 nests were located in beaver impoundments 6 to 14 acres in area. Stable water levels and tall, dense emergent plants apparently provide the necessary security, food supply, and nest support needed by these birds. Most preliminary nest-building is by the female, but the male helps gather nesting material and to a limited extent may assist in nest construction.

Females not only spend more time nest building but also are more effective in gathering materials (De Vos, 1964). De Vos did not observe the male actually incubating, but saw it sitting on the nest once during the egg-laying period. However, Griswold (1965) did report an instance of apparent incubation assistance by the male, inasmuch as both birds were once seen on the nest, with four eggs under one and three under the other. This is apparently the only report of possible incubation by the male, although a male sometimes stands over or sits on the nest when his mate is off foraging.

Of 74 completed clutches observed by Banko (1960), the average was 5.1 eggs, with a range of 3 to 9. Hansen et al. (1971) stated that 53 clutches from

Trumpeter swans, adult and first-winter juveniles

Alaska's Copper River area averaged 4.9 eggs, and 160 clutches from the Kenai region averaged 5.3 eggs. Yearly differences occur, with smaller clutches typical of years having late springs, and larger clutches typical of more favorable early breeding seasons. The eggs are laid at two-day intervals. Incubation period estimates have ranged from 32 to 37 days. Hansen et al. noted that six nests in the Copper River area had incubation periods of 33 to 35 days.

The cygnets normally all hatch at about the same time. However, Griswold (1965) reported a staggered hatching period in one captive pair. He noted that the first two young to hatch were seen entering the water initially when about 48 hours old, while the third left the nest when about 24 hours old. Griswold's observations were complicated by the fact that two females were present, and both may have contributed to the clutch.

Banko (1960) noted that hatching success of eggs of wild birds varied from 51 to 66 percent during three different years. During six years at Grande Prairie, Alberta, the comparable percentages ranged from 55 to 92 (Banko and Mackay, 1964), and three years' data from the Kenai Peninsula, Alaska, indicated an average 82 percent hatching success (Hansen et al., 1971).

Infertility and embryonic deaths appear to be the major causes of hatching failure, with egg predation insignificant. A few Alaskan nests have been found destroyed by bears (*Ursus*) and wolverines (*Gulo luscus*). Banko (1960) reported one probable instance of renesting following nest destruction. De Vos (1964) noted that for the first few weeks a youngster was closely guarded, with the two parents placing themselves on either side of the cygnet. However, the female was generally more closely associated with it, and usually when swimming the female led the cygnet, with the male following behind.

Griswold reported that by the age of about three months a captive juvenile female had attained a weight of 14.5 pounds. Four young males weighed from 13.5 to 16 pounds and collectively averaged about 15 pounds. Banko mentioned a 19-pound cygnet of preflight age, and Hansen et al. (1971) stated that such a weight can be attained in as few as 8 to 10 weeks.

Banko (1960) summarized data indicating that the fledging period is probably normally from 100 to 120 days, with known minimums and maximums of 91 and 122 days. A somewhat shorter range, from 90 to 105 days, has been reported for Alaskan birds (Hansen et al., 1971). Fall departure from the Alaska breeding grounds occurs in October, only shortly after the cygnets have fledged.

According to Banko, considerable prefledging mortality occurs, with possibly 50 percent or more of the young being lost during this period. Most of this mortality occurs early in life, from apparently varied causes. Monnie (1966) reported cygnet losses by great horned owls (*Bubo virginianus*) and probably also raccoons (*Procyon lotor*), and Banko suspected that minks (*Mustela vison*) or skunks (*Mephitis* spp.) might play a predatory role at the Red Rock Lakes NWR. Hansen et al. (1971) found a rather low (15–20 percent) cygnet mortality rate in Alaska over the first eight weeks after hatching, and practically none afterward.

Banko (1960) suspected that trumpeter swans are virtually free of most natural enemies once they have fledged and speculated that only coyotes (*Canis latrans*) or golden eagles (*Aquila chrysaetos*) might be of possible significance as predators. Females have been known to live as long as 18 years in the wild, and a male reached 23 years and 10 months (Baldassarre, 2014). The apparent maximum longevity record of the species in captivity is 32-plus years (Mitchell and Eichholz, 2019).

Some mortality occurs as a result of trumpeter swans being accidentally shot during legal whistling swan hunting seasons. Drewien et al. (1999) reported that 10 of 1,424 swans that were shot in Utah and identified by bill measurements were trumpeter swans, and 19 of 890 swans shot in Montana were judged by wing measurements to be trumpeter swans. Over the period 1994 to 2015, a total of 216 trumpeter swans were reported killed during tundra swan hunting seasons in Utah, Nevada, and Montana (Pacific Flyway Council, 2017).

Lead poisoning caused by the ingestion of lead shot or from other sources of lead, might be a more serious hunting-related problem, at least locally (Degernes, 1991; Wilson et al., 1998; Degernes et al., 2006; Smith et al., 2007). In a study of 400 swans (365 trumpeter, 35 whistling) that had died in northwestern Washington state in 2002 and were necropsied, 81 percent were lead-poisoned (Degernes et al., 2006).

Molts and migration

In Alaska, nonbreeding birds gather in flocks on large, open lakes and begin their wing molt almost simultaneously, with nearly all of them beginning and terminating their flightless period within ten days of one another. A less regular molting pattern occurs in breeding birds. In Montana, females usually begin their wing molt one to three weeks before the eggs hatch, while males of successfully breeding pairs may begin to molt either before or after hatching, but usually before. In Alaska, males usually begin their wing molt early in the incubation period, or sometimes as late as after hatching. There, females reportedly begin molting their flight feathers from 7 to 21 days after the clutch has hatched. Since the flightless period lasts about 30 days, both members of a pair are rarely flightless simultaneously, and both sexes regain their flying abilities prior to the fledging of the young. Molting in nonbreeding birds seems to be earlier and more synchronized than in breeders, but in either case the birds remain flightless for about 30 days (Hansen et al., 1971).

In Alaska, some young may still be unable to fly at the time of freeze-up, and these birds seem to postpone their fall migration as long as possible, with family groups being the last to leave the breeding grounds (Hansen et al., 1971). Starvation during severe winters may be a significant mortality factor, at least in Canada.

Mackay (1957) mentioned that cygnets of a family evidently remain together, or with their parents, for at least the first year after hatching, since three broodmates that had been banded in Alberta during 1955 were all illegally shot by a hunter in Nebraska the following fall. It is likely that, as in the Bewick's swan, social attachment to the parents persist for several years.

Trumpeter swans usually migrate only minimal distances to winter in areas that are free of ice and where food is plentiful. Thus, even some Minnesota flocks might move over distances of less than 25 miles to reach ice-free areas on the Mississippi River, but the majority of flocks in Wisconsin and Iowa

Trumpeter swans, adult and first-winter juveniles

migrate more than 400 miles to locations in states as far away as Kansas, Oklahoma, Texas, and Virginia, and especially to locations in Arkansas, Missouri, Illinois, and Indiana (Mitchell and Eichholz, 2019).

Status

The trumpeter swan was removed from the list of endangered species only after it became apparent that the Alaskan breeding population was much larger than previously known and that the overall population was thus appreciably greater than earlier estimated. Until 1946, when a small population of about 100 trumpeter swans was found in the Grand Prairie region on the British Columbia–Alberta border, it was thought that the only surviving populations of trumpeter swans were in the West Yellowstone region of Montana and Wyoming. The Alaska population was first detected during 1956 in the Copper River basin, and additional birds were found in the Kenai Peninsula the following year.

The earliest estimate of the Alaska population was by Hansen et al. (1971), who found nearly 3,000 swans in 1968, and probably missed seeing hundreds more during their aerial survey. There were also estimated to be about 150 swans that summer in western Canada, and over 300 in the vicinity of Red Rock Lakes NWR, Yellowstone National Park, and Grand Teton National Park. There were then about 600 additional birds in other US refuges and zoos, suggesting that about 3,500 existed in North America around 1970.

The first continental effort at estimating trumpeter swan numbers was the North America Trumpeter Swan Survey, which was initiated in 1968, and since 1975 has been repeated at five-year intervals. The estimated total numbers of "white swans" (adults and younger age classes with white body plumage) reported during those quinquennial surveys were as follows: 1968, 2,572; 1975, 3,727; 1980, 6,206; 1985, 8,812; 1990, 11,344; 1995, 14,809; 2000, 18,486; 2005, 25,006; 2010, 32,249; 2015, 76,016. The estimated overall average annual growth rates for adult and subadults were: 1968–1975, 7.7 percent; 1985–2000, 5.1 percent; 2000–2015, 6.3 percent. The long-term 1968–2005 rate was 6.6 percent (Groves, 2017).

The most rapid rates of population increase among white-plumaged swans for the entire 1968–2015 survey period occurred in the Interior Population (primarily the Nebraska and South Dakota portions of the "High Plains" population plus the Mississippi and Atlantic flyway regions) at an amazing 14.4 percent, followed by the Rocky Mountain Population (eastern British Columbia, Alberta, and Rocky Mountain states) at 6.5 percent, and the Pacific Coast Population (Alaska, western Yukon Territory and northwestern British Columbia) at 5.5 percent (Groves, 2017).

As these figures indicate, trumpeter swans have increased remarkably in North America since the 1970s. As an example of remarkable conservation success, the Pacific Coast Population, which winters from southern Alaska to Washington and Oregon, numbered 31,793 birds in 2015, of which 28,808 were in Alaska and 2,985 were in western Yukon Territory and northwestern British Columbia (Groves, 2017). The highest US Audubon Christmas Birds Counts in recent years have been at Skagit Bay, Washington, where they have rather consistently totaled about 2,500 to 3,500 birds. In Canada various sites in British Columbia, such as Chilliwack, Comox, Duncan, and Harrison River, have recorded recent Christmas Counts in excess of 1,000 birds. Several thousand birds of the Rocky Mountain Population usually overwinter at Red Rock Lakes NWR in Montana.

The Rocky Mountain Population had an estimated total of 17,164 swans in 2015, of which 16,143 were in Canada (eastern Yukon Territory, eastern British Columbia, Alberta, and Northwest Territories). There were also 300 in Wyoming (Yellowstone National Park, Snake and Green Rivers), 272 in Montana (Centennial, Madison and Paradise valleys, Blackfoot Indian Reservation), and 151 in Idaho (Island Park, Teton Basin, and Camus, Grays Lake and Bear Lake N.W.R.s). Smaller numbers were also present in central Oregon (Klamath Basin, Summer Lake and Malheur NWR) (24), and still smaller numbers in Washington (8), Nevada (2), and California (1) (Groves, 2017; Rocky Mountain Population Trumpeter Swan Subcommittee, 2017).

There were only two breeding pairs present in Yellowstone National Park in 2015. As I suggested earlier (1982, 2013), the abbreviated breeding seasons in mountainous locations with relatively short summers and frequent summer snows, such as Yellowstone and Grand Teton National Parks, and the extremely

Fig. 11. Trumpeter swan, adult landing

long fledging period (more than 100 days) of trumpeter swans that is required when breeding at these locations, makes effective breeding there only marginally successful, as has also been suggested by Schmidt et al. (2011) for Alaska.

Many of the Rocky Mountain Population birds breeding in Alberta, and probably also those breeding in Yukon and northwest British Columbia, winter with swans of the tristate segment of the US Rocky Mountain Population (Smith, 2013).

The Interior Population, which once bred across the Great Plains and eastward to the Great Lakes and might once have consisted of 100,000 birds (Gillette and Shea, 1995), was totally extirpated during settlement times but has effectively been recreated by transplant programs. These birds now breed from the high plains of Nebraska and South Dakota east to Michigan, Ontario, and New York.

Restoration efforts in Minnesota, Iowa, Ontario, Wisconsin, Ohio, and Michigan have resulted in many new nesting flocks within the Mississippi and Atlantic flyway components of the Interior Population. For example, Ontario's population began in 1992, beginning with about a dozen eggs taken from wild Alberta swans. In 2017 Indiana reported its first breeding of wild trumpeter swans. In Ohio 84 pairs produced 196 cygnets in 2019 from a restoration program that began in 1996. As of 2015 the combined populations of the Mississippi and Atlantic flyways numbered about 27,000 adult and subadult birds, plus an unknown number of cygnets.

The Interior Population's estimated 2015 swan population totaled 27,055 white-plumaged birds (plus an undetermined total number of juveniles), of which 17,021 were in Minnesota, 4,695 in Wisconsin, 3,021 in Michigan, and 1,471 in Ontario. There were also 311 in Nebraska, 204 in Iowa, 154 in Ohio, 97 in Manitoba, 54 in South Dakota, 22 in New York, and 3 in Pennsylvania (Groves, 2017).

As this eastern component of the Interior Population has expanded, the swans have pioneered south for the winter. Some are now wintering in wildlife refuges, sanctuaries, and other areas well to the south of the species' major breeding ranges (Burgess, 2002a). In Ontario a significant percentage of the province's swans winter in the La Salle Park harbor at Burlington, where they are fed by local volunteers. Many of the swans that nest in southern Minnesota fly only a short distance to winter on the Mississippi River (Linck, 1999). Wintering populations composed of migrant birds occur in eastern Kansas, northern Oklahoma, northern and central Missouri, northern Arkansas, southwestern Illinois, western West Virginia, and southern Ohio.

One of the best places to find wintering trumpeter swans in the Midwest is at Loess Bluffs NWR in northwestern Missouri. Nearly 2,000 trumpeter swans were present at Loess Bluffs during mid-December 2019, a population that has gradually increased over two decades. First noted in 2000, trumpeter swans have been present at this refuge with increasing regularity and numbers. The maximum number seen in 2008 was 78, but between 2009 and 2011 a peak average of 148 trumpeter swans were present (Johnsgard, 2012 [*Squaw Creek NWR*]). Since then the average numbers have increased about tenfold, with peaks usually occurring in early December and late February.

Other state conservation areas in Missouri that support wintering trumpeter swans include Four Rivers Conservation Area (CA), Columbia Bottoms CA, Cooley Lake CA, and Perry Memorial CA. One of the other areas with notably large winter concentrations is the National Audubon Society's Riverlands Migratory Bird Sanctuary near Alton in southwestern Illinois, where more than 1,000 trumpeter swans regularly overwinter. In Iowa, regular wintering sites include Dale Maffitt Reservoir near Des Moines, Hawkeye Wildlife Management Area in Johnson County, Lily Lake near Amana, and Beemer's Pond west of Fort Dodge.

Trumpeter swans also now regularly winter south to north-central Arkansas, such as at Heber Springs, Cleburne County, with some birds reaching northern Texas. The swans are rarely seen as far south as southern New Mexico and even more rarely into northern Mexico. Some of the more northerly wintering-only sites might develop into breeding locations in the future. Trumpeter swans are also increasingly venturing back to their probable historic Gulf Coast wintering areas in southern Louisiana and along the middle Atlantic Coast of eastern Virginia.

In 2015 breeding by the much smaller western (High Plains) component of the Interior Population was centered mostly in central Nebraska and southern South Dakota, plus a few isolated groups in eastern Manitoba, northern North Dakota, and eastern Wyoming and numbered more than 600 birds. Manitoba breeding probably developed from reintroduction programs in Ontario or the northern United States and was first documented in 2002 at Riding Mountain National Park. Nesting in Nebraska also developed gradually, as a result of incursions by the offspring of swans released during the early 1960s at South Dakota's Lacreek NWR. Since then nesting has expanded into many of Nebraska's Sandhills wetlands (Vrtiska and Comeau, 2009), especially in Cherry and Grant Counties, and has recently been reported from marshes in at least five other Sandhills counties.

However, South Dakota's swan population has declined as Nebraska's has increased, and most of this restored flock now both nests and winters in Nebraska. Breeding sites have expanded to include some locations outside the Sandhills region, and Nebraska's Audubon Christmas Bird Count numbers have gradually increased, reaching 206 birds by the 2018–19 count. By 2019

Trumpeter swans, adults in flight

wintering was common on the Snake, North Loup, and North Platte Rivers, with flocks of 100 or more birds sometimes reported.

During the course of reintroduction efforts, a variety of identification methods have been used to field-verify origins of released birds, including neck collars, wing tags, and foot bands, and by using combinations of band and tag colors, and incorporating alpha-numeric codes to provide more specific geographic origin and date identifications. Protocols that have been established for interpreting these identification codes on trumpeter and tundra swans have been summarized by the Trumpeter Swan Society (2010) and might be of interest when viewing collars, bands, or tags on wintering birds.

The long-term (1966–2017) continental population trends for trumpeter swans reported during the annual Audubon Christmas Bird Counts as of 2017 (https://www.audubon.org/conservation/where-have-all-birds-gone) have indicated spectacular annual increases of 10.28 percent in the United States and 5.68 percent in Canada (Meehan et al., 2018).

In summary, the 2015 continental trumpeter swan population, exclusive of cygnets, was estimated as consisting of 31,793 birds in the Pacific Coast population, 17,164 in the Rocky Mountain Population, and 27,055 in the Interior Population for a total of about 76,000 birds (Groves, 2017). If, as in whistling swans, juveniles typically make up about 12 percent of winter populations (Caswell et al., 2007), about 9,000 additional first-year birds might be assumed to have also been present.

Whooper Swan *Cygnus cygnus* (Linneaus) 1758

Other vernacular names

None in general English use. Singschwan (German); cygne sauvage (French); cisne gritón (Spanish).

Subspecies

No subspecies is currently recognized. Rare in western North America; most common in the Aleutian Islands—has bred at least twice on Attu Island (Kenyon, 1963). Also rare in mainland Alaska and increasingly rarer along the Pacific Coast southward during migration or winter periods, with sight records south to California. Numerous Atlantic Coast sight records extend south to Florida (see the Range section); possibly all are the result of escapes from captivity, but Icelandic vagrants are possible. No breedings of proven wild birds have yet been reported from mainland North America, although breedings by apparently feral pairs have occurred.

Range

The species' primary breeding habitat is the taiga and scrub forest zone. Its western Eurasian breeding range includes Iceland and northern Eurasia, from northeastern Finland and European Russia east to west-central Russia in the Ural Mountains. A geographically much larger eastern population extends across Asia from the Ural Mountains east to the Kamchatka Peninsula and south along the Sea of Okhotsk coast to the lower Amur River valley (about 52°N latitude). Insular populations breed on the Commander Islands and Sakhalin Island. The whooper swan also previously bred in Greenland, where it is now a rare vagrant.

The Icelandic and western Euro-Asian population winters from Great Britain and northwestern Europe (especially Germany, Denmark, and Sweden) east perhaps to the Black and Caspian Seas, although data on the Euro-Asian population's eastern limits are apparently lacking. The central and eastern Asian populations winter in relatively small numbers (probably no more than a few thousand) in Armenia, Kazakhstan, Uzbekistan, Kirgizstan, and Mongolia. Mostly larger numbers winter in Japan (more than 30,000), China (about 10,000–15,000), and South Korea (probably less than 2,000), with less certain numbers for Kamchatka and Sakhalin Islands, as of the early 2000s (Brazil, 2003). Small numbers also regularly winter in the Aleutian Islands.

Map 3. Breeding (hatched) and wintering (stippled) distributions of the whooper swan (adapted from Johnsgard, 1978 and Brazil, 2003). The lower inset is the Aleutian Islands, indicating sighting records (small arrowheads) and Attu breeding records (large arrowhead). See text for island names.

Measurements and weights

Wing: Males 590–640 mm (23.6–25.2 in); females 581–609 mm (22.9–24.0 in). *Culmen:* Males 102–116 mm (18.7–28 in); females 97–112 mm (3.8–4.4 in) (Scott and the Wildfowl Trust, 1972). Brazil (2003) provided large, both-sexes samples of wing, skull, and tarsal measurements for adults, yearlings, and juveniles from the UK and Japan. *Eggs:* Average 113 × 73 mm, white, 330 g.

Weights: Males 8.5–12.7 kg (18.7–28.0 lb), average 10.8 kg (23.8 lb); females 7.5–8.7 kg (16.5–19.2 lb); average 8.1 kg (17.8 lb) (Scott and the Wildfowl Trust, 1972). Brazil (2003) provided extensive both-sexes winter weight data of adults, yearlings, and juveniles from the UK, both-sexes adult and juvenile winter weights from Denmark and Japan, and both-sexes summer weights of adults from Iceland and Finland.

Identification and field marks

In the hand: Length 55–65 inches (140–65 cm). The whooper is the only one of the northern swans that has yellow on its bill reaching forward beyond the nostrils and the only swan with a partially yellow bill that has a wing length in excess of 580 millimeters. The culmen length (95–112 mm) is slightly longer than that of a tundra swan and slightly shorter than that of a trumpeter swan. In the adult plumage both sexes are entirely white, with black legs and feet and a black bill except for the large yellow bill marking. Females cannot be externally distinguished from males. Juveniles have a variable number of grayish feathers mixed with the white feathers of the second-year plumage, and their bills are pinkish at the base rather than yellow.

In the field: The highly extensive area of yellow on its bill is the whooper swan's best field mark. The relative amounts of yellow on the bills of both yearlings and adults should distinguish them from Bewick's and whistling swans; in whooper swans the yellow extends beyond the nostrils, whereas in the Bewick's and whistling swans it terminates behind the nostrils. The more trumpetlike and less musical call of the whooper swan helps to distinguish it from the smaller whistling and Bewick's swans. Field separation of Bewick's swans from whooper swans is unlikely to be an issue in North America, except perhaps in the Aleutian Islands. In flight, the Bewick's swan has a shorter-appearing neck and a more rounded head, and the voice of the Bewick's swan is higher in pitch and more musical than that of the whooper swan (Rees, 2006). The more trumpetlike whooper swan's calls are typically double-noted, like a common call of the trumpeter swan, with the second note higher in pitch (Brazil, 2003). The species is capable of maintaining a precisely timed antiphonal duet series, resulting in a sequence between pair members that gradually becomes more synchronized and might contain as many as 15 coordinated vocal elements (Hall-Craggs, 1974).

Evolutionary Relationships

Although the whooper swan and trumpeter swan have at times been regarded as conspecific (e.g., Delacour, 1954–64), the usual present taxonomic interpretation is to consider the two as separate species and presumably each other's nearest relative. However, Parkes (1958) and Livezey (1996) judged the Bewick's swan, rather than the trumpeter swan, to be the whooper swan's nearest living relative.

Natural History

Habitats

The preferred breeding habitat of whooper swans is shallow freshwater pools and lakes and along slowly flowing rivers, primarily in the coniferous forest (taiga) zone but also in the birch forest zone and treeless plateaus, rarely in tundra (Voous, 1960). Brazil (2003) stated that the species' habitats vary over the collective range of the species, but the birds are most often seen on fresh water, or at least wetland habitats that range from shallow to deep water, deep lakes to shallow lagoons, marshes and flooded meadows. During the breeding season it favors boreal forest or taiga, with freshwater lakes, small pools, slow-moving rivers, marshes, and swampy locations for nesting. The breeding location should have sufficient water as to allow for aquatic takeoffs and should remain ice-free over an entire breeding season of at least 130 days. The chosen nest site is often very close to water and might be on a small island or headland, a river bank, or on an elevated area close to water. The breeding territory should be large enough to provide food for the entire family, which might be an entire lake or a part of a smaller, more prolific area of marshland.

Foods and foraging

On the breeding grounds the foods of adult whooper swans probably consist mainly of the leaves, stems, and roots of aquatic plants, including algae. A considerable amount of grazing on shoreline and terrestrial vegetation is performed; the whooper swan is thought to consume higher proportions of terrestrial plants and animal materials than is true of the Bewick's swan. In the wintering areas of Europe the birds have been found to consume such aquatic plants as pondweeds and similar leafy foods (*Zostera*, *Ruppia*, *Elodea*) that can readily be reached from the surface. In some areas they also graze on winter wheat, waste grain, turnips, and potatoes (Owen and Kear, in Scott and the Wildfowl Trust, 1972). Animal materials, such as midges and freshwater mussels, are consumed in some areas as well (Kear, 2005).

Plants that have been found in diverse parts of the whooper swan's range include aquatic eelgrasses (*Zostera* spp.), wigeon grass (*Ruppia maritima*),

Whooper swans, adult and first-year juveniles

pondweeds (*Potamogeton* spp.), Canadian pondweed (*Elodea canadensis*), and stonewort algae (*Characeae*), plus terrestrial sweet-grasses (*Glyceria* spp.). Other terrestrial grass taxa, such as bent grass (*Agrostis*) and marsh foxtail (*Alopecurus*), are often important foods when the birds forage in meadows.

Brazil (2003) provided an exhaustive analysis of whooper swan foods, as well as their foraging methods, the development of field foraging during the later twentieth century, foraging associations with other waterfowl species, and the influence of temperature on feeding behavior. He tabulated a wide array of foods reportedly eaten by whooper swans, including more than 40 taxa (mostly species and genera) of aquatic and marshland plants, 30 taxa of terrestrial noncrop plants (the majority consisting of 11 grass and 5 sedge taxa), and 8 taxa of crop plants (including varied grasses, sugar beets, turnips, barley, wheat, rye, potatoes, and rapeseed). A number of other more generalized plant materials and some animal taxa have also been reported as foods, including aquatic insects (which are especially eaten by cygnets), mollusks, fish eggs, frogs, and other minor or miscellaneous items.

While foraging on water, whooper swans typically ingest food by dabbling the bill at or just below the surface or by dipping the head or head and neck below the surface, often in association with foot paddling movements that loosen plant roots and stir up food items from the substrate. In deeper water the birds upend (also called tipping-up) so that the entire front half of the body is submerged. When feeding on land, the birds might pick food from the surface, graze, dig into soft soil with the bill, shovel surface foods into the lower mandible by using it as a scoop, or peck at larger food items, using the nail of the bill or the edges of the mandibles to cut or break off food particles small enough to swallow (Brazil, 2003).

Like trumpeter swans and tundra swans, whooper swans have increasingly resorted to consuming terrestrial human-planted foods in recent decades, in part because their historic winter maritime and other aquatic food sources have mostly declined or disappeared. In Britain and elsewhere in Europe, agricultural crops, such as potatoes, turnips, and sugar beets, as well as grasses and waste grain, have become significant sources of carbohydrates for wintering whooper swans.

Social Behavior

Winter social behavior

Whooper swans are gregarious and form large flocks during the nonbreeding season. In Aberdeenshire, Scotland, flocks of 300 to 400 swans often may be seen foraging in agricultural fields. Fairly large winter flocks also occur locally in Japan at protected sites such as Lake Hyoko reservoir, Honshu. These

Whooper swan, adult head profile

wintering flocks consist of firmly paired, courting, and sexually immature cohorts, including juveniles still closely consorting with their parents.

First-year birds remain with their parents through the winter period and start back toward the breeding grounds with them. Whooper swans leave their European and Asian wintering grounds in late February or March and may not arrive at their northernmost breeding areas until May. Since the birds arrive on the breeding grounds in pairs, family social affiliations must be broken during the spring migration period, and some mate changes might also occur.

Reproductive Biology

Sexual behavior and pair-bonding

So far as is known, the process of pair formation and pair-bonding in whooper swans is very similar to that of the trumpeter and tundra swans, as is their behavior associated with copulation.

Black and Reese (1984) reported that, among wintering birds, some single males and females that became firmly paired by their second winter were with their same mates two winters later, but pairing was much more frequent by birds in their third winter, especially among females. By their fourth winter, 14 of 16 known-age females were paired and presumptive potential breeders. Five pairs banded in 1979–80 were still paired in 1983–84. In the wild, initial first pairing occurs at two to three years of age, and first breeding occurs at four to six years, whereas in captivity sexual maturity is reached in two to three years and pair-bonds have lasted 4 to 13 years in that situation. In the wild, the longest known duration of a pair-bond is 11 years (Rees, 2006).

Brazil (2003) reported (based on two observations) that copulatory behavior differs from that of the trumpeter in that the pre-copulatory actions were more obviously bathing-like and were more prolonged than Johnsgard (1965) had observed; the post-copulatory display by the male did not involve wing-spreading.

Although essentially monogamous, divorce rates are much higher in whooper swans than in Bewick's swans. In one study 5.8 percent of the whooper pairs studied were found to have a new mate the following year, even though their earlier mate was still alive. Divorce is not obviously associated with poor breeding success, since 25 percent of divorced pairs had bred successfully the previous year. However, breeding success improved among pairs remaining together for several years (Rees et al. 1996). Polygamous pair-bonding has been reported at least once in the wild and twice in captivity (Mitchell and Eichholz, 2019).

Breeding biology

Immediately upon the birds' arrival on their breeding grounds, or at most within two weeks of arrival, nest building begins. Nests are built either on dry ground or in reed beds, often so large and with so deep a cup that the top of the sitting female may be flush with the rim of the nest. The nest cup is extensively lined with down. Eggs are laid at approximately 48-hour intervals, and a clutch of from four to seven eggs, most commonly five, is laid. Average clutch sizes in Iceland and Finland were 4.5 and 4.4, respectively, with clutches in southern Finland slightly larger (and cygnet mortality lower) than in northern Finland (Einarsson, 1996). Clutch sizes vary with the productivity of the breeding habitat; those locations having more available food or better food quality are associated with larger clutches.

In Russia, egg-laying occurs during May and June, and in Iceland it also normally occurs at that time. Undoubtedly the time at which the nesting sites become snow-free dictates the year-to-year onset of nesting in this species. The probable usual incubation period is 35 days, although various estimates have ranged widely, from 31 to 42 days.

Incubation is performed entirely by the female, but the male remains in very close attendance. Typically she leaves for a short time during the warmest part of each day to forage. Brazil (2003) noted that the tundra-breeding whistling and Bewick's swans minimize their breeding season durations because males incubate during times the female is absent from the nest, resulting in incubation constancy rates of 97 to 99 percent (the eggs left unheated only 1 to 3 percent of the time), whereas in the whooper swan only the female incubates.

Fig. 12. Whooper swans, adults landing

Fig. 13. Whooper swans, pair performing triumph ceremony

In the trumpeter swan there is a report of one male nest-sitting (Henson and Cooper, 1992), and the male might incubate as much as 1.0 to 1.75 percent of the time (Mitchell and Eichholz, 2019).

A few cases of second clutches being laid following the loss of the first clutch have been reported, the second clutch being laid 19 to 23 days later and having fewer eggs. Hatching of the clutch is relatively synchronous, occurring over a 36- to 48-hour period. The male also closely guards the cygnets, which in

Whooper swans, adults with downy cygnets

Iceland have been reported to fledge in as short a period as about two months but is typically 87 days, with a range of 78 to 96 days (Haapenen, Helminen, and Soumalainen, 1973b). In China and Russia the fledging period of wild birds has been judged to be somewhat longer, at 80 to 105 days (Brazil, 2003). Captive-raised young may fledge in as little as 80 days, with a range of 77 to 84 days (Kear, 2005) as compared with 60 to 70 days in the Bewick's swan (Rees, 2006). The young birds eat insect larvae, adult insects, and vegetation growing on the water surface or just below it. The juveniles remain with their parents through at least their first winter.

The average percentage of young that hatch and survive to reach wintering grounds is highly variable from year to year and place to place. Brazil (2003)

Whooper swans, adults with downy cygnets

summarized data on juvenile percentages for 11 different regions and two different years. A total 16,742 birds were aged in 1986 on wintering areas, and the proportions of young varied from 17.1 percent to 26.9 percent. In 1991, with a sample count of 18,035 birds, percentages varied from 5.9 percent to 18.6 percent among three locations. In Britain and Ireland, long-term counts averaged 19.6 percent young for the years 1948 to 1984.

A set of brood size counts taken in five different wintering areas of the British Isles and Iceland that involved counts of 1,805 broods in 1991 revealed the mean frequencies of occurrence of the following observed brood sizes: (1) 27.7 percent, (2) 36.2 percent, (3) 22.6 percent, (4) 12.5 percent, and (5) 1.0 percent. In 1995 a survey of 10,156 birds indicated that 17.9 percent were cygnets, and the overall mean brood size was 2.32 young. The highest proportion of young in winter populations reported was that for Ireland's in 1985, with 26 percent young, although from 1948 to 1984 the overall Icelandic mean was 19.6 percent (Brazil, 2003).

Varner and Eichholz (2012) calculated annual survival rates for all age classes to range from 74.9 percent to 79.1 percent for short-distance migrants receiving food supplements on wintering areas, and 82.3 to 86.3 percent for long-distance migrants living on aquatic vegetation and field crops.

Molting and migration

During early summer to midsummer subadult and adult whooper swans molt their flight feathers (primaries and secondaries) and become flightless for a period of about a month. In whooper and trumpeter populations the overall molting period lasts about four to six weeks and extends from early June to early September. Both sexes of nonbreeding whooper and trumpeter swans molt their feathers simultaneously, but in breeding whooper swans and mute swans females begin molting first. In Iceland the differential timing of the sexes in breeding birds is such that one member of the pair is capable of flight through nearly the entire summer (Brazil, 2003), and in trumpeter swans the molting period of breeding birds is similarly extended to nearly twice the length of nonbreeders. Juveniles gradually undergo complete molts that begin with losing their gray body feathers during their first fall and are completed the next summer with wing molts like those of adults, by which time they cannot be visually distinguished from them.

Most studies suggest that the most hazardous period in a swan's life is from hatching through fledging, and especially during the first few weeks of life, as summarized by Brazil (2003). In Finland, Bart, Earnst, and Bacon (1991) found that the percentage of eggs surviving from laying to hatching was 74 percent, that 90 percent of the cygnets survived from hatching to fledging, and that overall survival from eggs to fledging was 59 percent. Knudsen, Laubeck, and Ohtonen (2002) reported that the survival rates of cygnets to wintering grounds varied from 51.9 percent for those raised in peatlands to 76.4 percent for those raised in eutrophic lakes; the latter group also averaged substantially heavier in body mass than cygnets raised on peatlands or oligotrophic lakes. Various studies suggest that the fall migration period is not particularly hazardous to young birds (Gardarsson and Skarphedinsson, 1984; Rees et al., 1991).

An annual survival rate of 85.1 percent has been calculated for British whooper swans, as compared with 81.8 percent for mute swans and more than 89 percent for Bewick's swans (Brown, Linton, and Rees, 1992). Haapenen (1991) calculated that the survival rate of first-year birds in Finland is about 30 percent, of second-year birds is 75 percent, and of older birds is 88 percent.

Whooper swans, adults with a first-winter juvenile

The maximum longevity record for whooper swans in captivity is at least 25 years (Scott and the Wildfowl Trust, 1972), and the oldest known wild bird record is 22-plus years (Brazil, 2003).

Status

In Alaska, whooper swans are seasonally uncommon to rare in southwestern Alaska (Alaska Peninsula and Aleutian Islands), where they are mostly found in the central and western islands and bred on Attu Island in 1996 and 1997, and where a few dozen winter. More rarely they occur in the Pribilof Islands before early May (Armstrong, 2015) as well as St. Lawrence Island. McEneaney (2004) reported records of single birds or small groups from the outer Aleutians, including (west to east) Attu, Shemya, Buldir, Little Kiska, Amchitka, and Adak.

McEneaney (2004) also summarized available data to early 2004 on mainland North American whooper swan occurrences, exclusive of Greenland (where it is a rare vagrant but once bred). He tallied nearly 60 records from eastern North America, and 30 records from western North America, exclusive of Alaska. Apart from observations in the Pacific coastal states and provinces of California (16), British Columbia (5) and Oregon (3), there were also records from Alberta (2). Ontario (1) Montana (2) and Wyoming (1).

Many whooper swan records from noncoastal areas could have been the result of escapes by captive birds, especially those from the eastern states, where many aviculturists maintain and have sometimes bred captive whooper swans. Coastal eastern states and provinces with recorded occurrences (to early 2004) include Maine (7), Quebec (5), Massachusetts (5), Maryland (2), New Brunswick (2), New York (2), Labrador (1), New Jersey (2), North Carolina (1) and Florida (1). Interior states and provinces with records include Minnesota (6), New Hampshire (5), Ontario (4), Iowa (2), North Dakota (2), Ohio (1), and Iowa (1) (McEneaney, 2004).

There have been at least two records of whooper swans breeding in the eastern states. In 1966 and 1968 they bred on a small farm near Ipswich, producing four and five young, respectively. Breeding also occurred in Minnesota in 1998, where an adult of unknown history was found with four young in Washington County. Such interior breedings are most likely to have been the result of released or escaped captive birds (McEneaney, 2004).

The total world population of the whooper swan was less than 100,000 birds in about 1970, but nearly universal protection has been given to the species. The Icelandic breeding component was then 5,000 to 6,000 birds, which wintered in Iceland and Great Britain. Those breeding in Scandinavia and western Russia and wintering in northwestern Europe probably numbered about 14,000. The birds that bred farther east in Russia wintered mainly on the Black and Caspian Seas and probably then totaled at least 25,000. The Far Eastern breeding component that winters along the western Pacific coast was least well documented, but close to 11,000 were counted during a wintering census in Japan during that same time period (Scott and the Wildfowl Trust, 1972).

In more recent surveys, the Icelandic population was judged to be nearly 21,000 in a 2000 survey, with 30 to 33 percent wintering in Great Britain, 61 to 66 percent in Ireland, and the rest wintering in Iceland (Kear, 2005). The northwestern European winter population was surveyed in January 1995, with over 20,000 birds judged to be present in Denmark, 14,000 in Germany, 7,500 in Sweden, more than 5,000 in Norway, and 3,000 in Poland. The total

Whooper swans, adults with first-winter juveniles

northwestern European population was estimated at 59,000 in the late 1990s (Laubek et al., 1999). The whooper swans breeding in Finland, Sweden, and Norway winter along the coasts of Norway and southern Sweden as well as in Denmark, Germany, and elsewhere in western Europe. There is a small but increasing winter population in Greece and Turkey as well, perhaps emanating from the Caspian area (Rees et al., 2019).

The species' eastern population is dispersed across Russia and northern China, with those breeding in western Russia east to the Ural Mountains and wintering around the Black and Caspian Seas of uncertain population size, but possibly numbered about 14,000 to 20,000 in the 1970s. Those western, central, and eastern Siberian swans breeding from the Ural Mountains across Russia to the Russian Far East and wintering in China, Japan, and Korea are likewise of uncertain population size, but summer surveys of the West Siberian plain have resulted in very high estimates ranging from 252,000 to 380,000 birds (Brazil, 2003). No winter surveys are available to confirm such large breeding season numbers.

During the 1990s there were winter estimates of about 60,000 in China and 31,000 in Japan. Perhaps another 4,000 birds wintered on the Korean peninsula, and a few also wintered on the Aleutian and Pribilof Islands (Kear, 2005). At present there seems to be no real threat to this species' existence, at least on an overall basis.

Brazil (2003) estimated the world population at the turn of the twenty-first century to be between 150,000 and 200,000 birds, making it probably the most numerous of the northern swans. The most recent estimates of the Asian nonbreeding season population of the whooper swan are from Qiang Jia et al. (2016). Based on combined counts from Korea, Japan, and China, the estimated total population was 42,000 to 47,000 swans. Rees et al. (2019) provided a recent collective world population estimate of about 263,000 birds, with 56 percent of the total in the northwest mainland Europe population.

Tundra Swan (Whistling and Bewick's Swans)
Cygnus columbianus (Ord) 1815

Other vernacular names

Whistling swan: Pfeifschwan (German); cygne siffleur (French), cisne silbador (Spanish).

Bewick's swan: Zwergschwan (German); cygne de Bewick's (French); cisne de Bewick's (Spanish).

Subspecies and range

Cygnus columbianus columbianus (whistling swan):

Breeds in Arctic tundra habitats from western Alaska to Hudson Bay including Yukon, Northwest Territories, northeastern Manitoba, northern Ontario, and northwestern Newfoundland (Labrador), and on Southampton, Banks, Victoria, and St. Lawrence Islands. The densest populations occur in the Yukon-Kuskokwim deltas of Alaska and the Mackenzie and Anderson River deltas of Northwest Territories (Limpert and Earnst, 1994). Since 1974 whistling swans have been breeding in extreme northeastern Siberia (Chukotka Peninsula), from the Bering Sea coast (192°E longitude) east along the Chukchi Sea coast to Nolde Bay (174°E longitude) and south along Anadyr Bay possibly to Kresta Bay (65°30'N latitude). It is possible that, as of 2002, some 600 to 1,000 birds might be summering in the Chukotka Peninsula. There, their distribution overlaps with that of the Bewick's swan, and mixed pairs as well as probable hybrids have been seen (Evans and Sladen 1980; Rees, 2006).

Winter: The Western Population of *Cygnus c. columbianus* winters in small numbers along the Pacific coast from the Aleutian Islands and south coast of Alaska but mainly concentrates from Vancouver Island, British Columbia, south through coastal Washington to the Central Valley of California, with far fewer numbers wintering as far south as the lower Colorado River valley. Inland wintering also occurs locally in southern British Columbia, southwestern Oregon, western Nevada, northern Utah, southern Idaho, and western Montana. About three-fourths of the Western Population passes through Utah (Bear River marshes and Great Salt Lake region) while on migration, and varying numbers often overwinter there.

The Eastern Population of *columbianus* predominantly winters on the Atlantic coast, mainly from New Jersey south through Chesapeake Bay to coastal

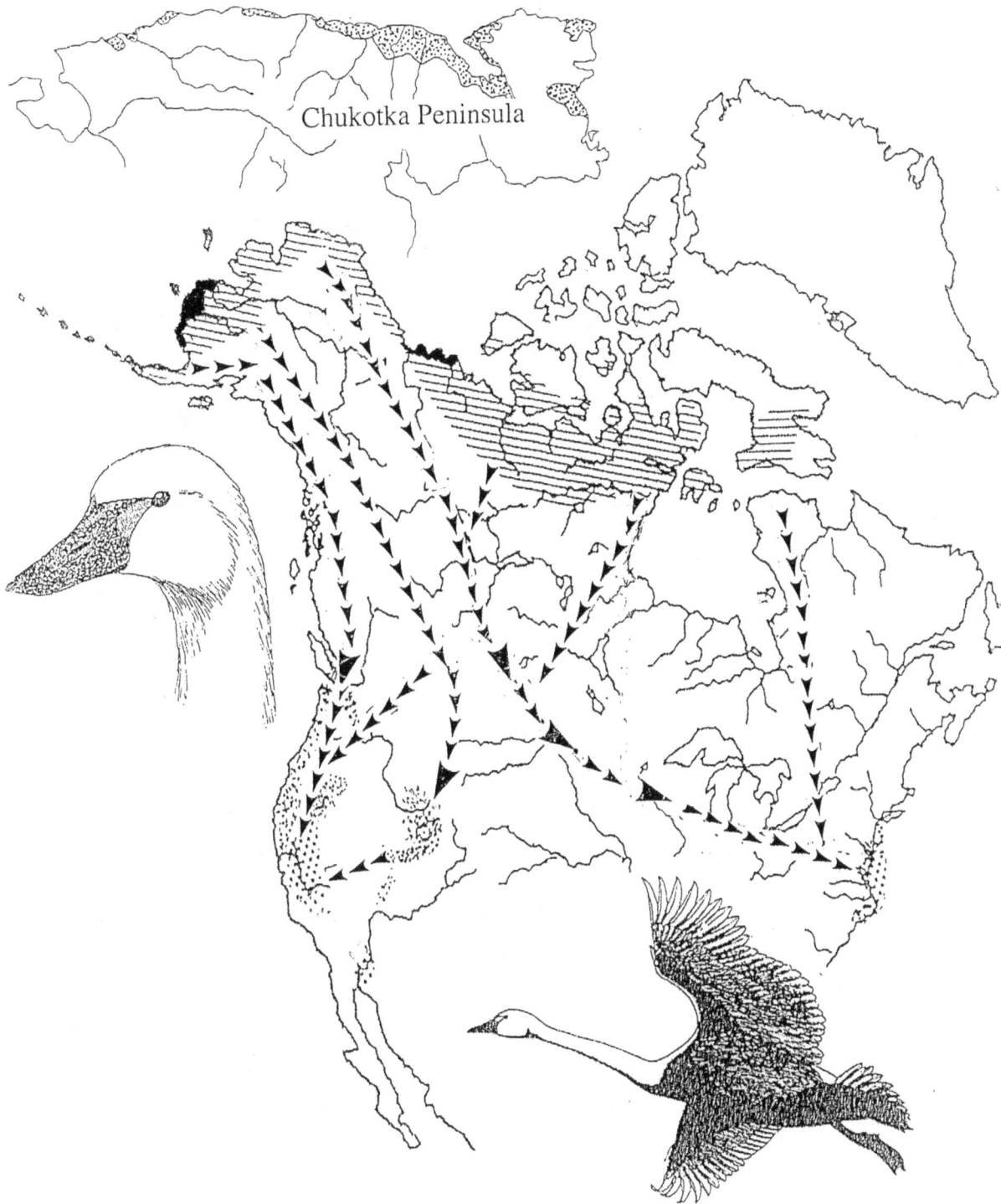

Map 4. Breeding (hatched) and wintering (stippled) distributions of the whistling (tundra) swan. Regions of densest breeding are inked. Arrowheads indicate major whistling swan migratory routes, with larger arrowheads showing important staging areas. This taxon's summer distribution on Russia's Chukotka Peninsula is also shown (upper inset), with known breeding areas as of 2002 indicated with stippling (from Rees, 2006).

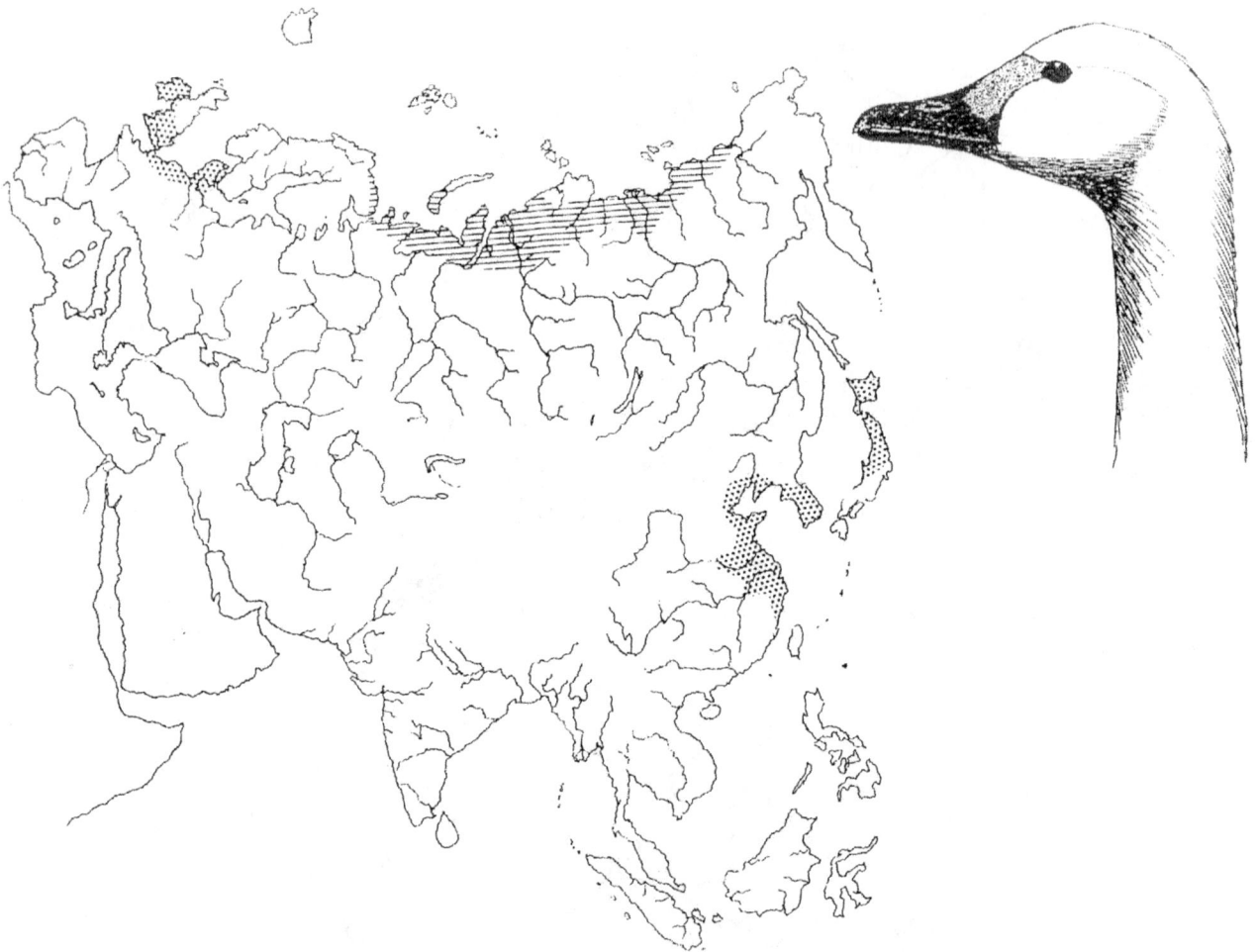

Map 5. Breeding (hatched) and wintering (stippled) distributions of Bewick's (tundra) swan (adapted from Johnsgard, 1978 and Rees, 2006). The summer distribution of the whistling swan taxon on Russia's Chukotka Peninsula is indicated by stippling in Map 4.

North Carolina. Some wintering occurs south to coastal South Carolina and north to coastal New England. Limited wintering also occurs in the eastern Great Lakes region and south to the coasts of Texas and Louisiana (Limpert and Earnst, 1994).

Cygnus columbianus bewickii (Bewick's swan):

Breeds in Russia from the Pechenga River, near the Fenno-Russian border (31°E longitude), eastward along the north Siberian coast to the Kolyma River delta and Chaun Bay (Chaunskaya Guba), to Chukotka (eastern Yakutia) at about 170°E longitude (Rees, 2006). Breeding has also been reported on Wrangel Island, Kolguev Island, and southern Novaya Zemlya. An eastern Siberian race, *jankowskii*, has sometimes been attributed to birds breeding east of the Lena River delta (126°E longitude), but its validity is doubtful (Rees, 2006).

Winter: The Bewick's swan winters in several widely separated areas. The population that breeds in European Russia winters in Europe (especially the Netherlands). Small numbers of those swans that breed directly west of the Ural Mountains (the western Siberian breeding population) winter along the Caspian Sea, while most, as well as those breeding in central Siberia (the western Siberian breeding population), winter in eastern Asia (China, Japan, and Korea). Small numbers of Eurasian-breeding vagrants have been seen north to Iceland, Bear Island, and Svalbard. North American vagrants have been seen in Alaska, Alberta, Saskatchewan, Oregon, and California.

Measurements and weights

Whistling swan (Cygnus columbianus columbianus):

Measurements: Wing: Adult male 480–574 mm (avg. 533 mm), adult female 505–561 mm (avg. 531.6 mm). *Culmen:* Adult male 81–110 mm (avg. 95 mm), adult female 79–102 mm (avg. 91 mm) (Rees, 2006). *Wing:* Adult male 501–569 mm (avg. 538 mm), adult female 505–561 mm (avg. 531.6 mm). *Culmen:* Adult male 97–107 mm (avg. 102.6 mm), adult female 92.5–106 mm (avg. 99.9 mm) (Scott and the Wildfowl Trust, 1972). *Eggs:* Average 110 × 73 mm, white, 280 g (9.9 oz).

Weights: Adult male 4.7–9.6 kg (10.4–21.2 lb), average 7.1 kg (15.6 lb); adult female 4.3–8.2 kg (8.8–18.1 lb); average 6.2 kg (13.7 lb) (Scott and the Wildfowl Trust, 1972). Banko (1960) reported that seven males at least two years of age had a maximum weight of 19.5 lb (8.8 kg), and 21 females of the same age class had a maximum weight of 19.0 lb (8.6 kg). Nelson and Martin (1953) indicated an average weight of 35 males as 15.8 lb (7.2 kg), with a maximum of 18.6 lb (8.4 kg); 42 females averaged 13.6 lb (6.2 kg), with a maximum of 18.3 lb (8.3 kg).

Bewick's swan (Cygnus columbianus bewickii):

Measurements: Wing: Adult male 485–573 mm; adult female 478–543 mm. *Culmen:* Adult male 82–108 mm; adult female 75–100 mm (Scott and the Wildfowl Trust, 1972).

Weights: Adult male 4.9–7.8 kg (10.8–17.2 lb), average 6.4 kg (14.1 lb); adult female 3.4–6.4 kg (7.5–14.1 lb), average 5.0 kg (11.0 lb) (Scott and the Wildfowl Trust, 1972). *Eggs:* Average 103 × 67 mm, white, 260 g (9.2 oz.). Rees (2006) compared weights and measurements (wing, skull, bill, and tarsus) of large samples of Bewick's and whistling swans, with both-sexes data for adults, yearlings, and cygnets of the Bewick's swan, and comparable data for weight, bill, and tarsal measurements of the whistling swan. The percentage positive difference in mean mass for adult male whistling swans over Bewick's swans (n = 1,447 vs. 1,201) was 9.1 percent, for adult females (n = 1290 vs. 946) was 6.7 percent, for yearling males (n = 87 vs. 206) was 4.7 percent, and for yearling females (n = 57 vs. 170) was 2.7 percent. The percentage positive difference in mean wing length for adult male whistling swans over Bewick's swans (n = 8 vs. 512) was 0.5 percent and for adult females (n = 15 vs. 414) was 3.7 percent.

Identification and field marks

Tundra swan (both subspecies)

In the hand: No external differences in the sexes of tundra swans exist that would allow for sex determination without internal examination. Birds possessing feathered lores and/or some grayish feathers persisting from the juvenile plumage are in their first year of life. Apparently the rate of sternal penetration of the trachea is fairly constant for the first three years, and by the second winter the tracheal loop starts to rotate and begin its expansion into the keel of the sternum (Tate, 1966). First-year birds have a well-defined "V" groove formed by the nasal and lachrymal bones, which gradually alters by medial fusion with age, so that the V is nearly obliterated in old birds. In young birds the feathers of the forehead extend forward to a point in the midline, while in older birds this point gradually recedes until a smooth and rounded brow is formed. Together with the length of the tracheal perimeter within the sternum, the changes in the shape of the nasal bones are good indicators of age in tundra swans, according to Tate.

Whistling swan

In the hand: Length 48–58 inches (120–150 cm). The whistling swan is completely white in adult plumage, with black legs and feet and a bill that is typically entirely black except for a small yellow area in front of the eye. However, some whistling swans lack this yellow mark, and thus a bill length that is less than 50 mm from the front of the nostrils to the tip of the bill is a better criterion for birds in the hand. Females are identical to males and average slightly smaller in size. Juveniles possess some gray feathers for most of their first fall and winter of life, and their bills are mostly pinkish.

The whistling swan is somewhat smaller than a trumpeter swan (overall tail-to-bill length 48–58 inches [120–150 cm]), and its bill too is typically entirely black. To be certain of identification, the upper surface of the sternum must be examined to see if a dorsal protrusion near its anterior end is present, which

Fig. 14. Whistling swans, pair in flight

would indicate a trumpeter swan (see Fig. 9). Alternatively, the bird is probably a tundra swan if it weighs less than 20 pounds, measures less than 50 mm from the tip of the bill to the anterior end of the nostril, and has bright yellow or orange-yellow spots on the lores and the basal portion of the upper mandible. There are occasional exceptions to this trait; in Idaho 2 of 698 trumpeter swans were found to have yellow lores, whereas in Maryland about 10 percent of the whistling swans banded there lacked yellow lores (Baldassarre, 2014). In another study of eastern whistling swans 3 percent of the birds lacked yellow markings (Evans and Sladen, 1980). Drewien and Bouffard (1994) provided data on both species' mensural traits, including sexual and age differences in various bill, toe, and tarsal measurements, as described in the trumpeter swan account.

In the field: Unless both trumpeter and tundra swans are seen together, a size criterion is of little value in the field for distinguishing them. The neck of tundra swans is often noticeably shorter in flying birds, and the bill profile often is somewhat more concave than that of the trumpeter swan, especially in adult males. The amount of bare skin in front of the eyes is less wide in tundra swans than in trumpeter swan, so that at any distance the eye is less evident in trumpeter swans. Although the extensive yellow bill markings of the Bewick's swan are a reliable field mark for identifying birds of that race, in the whistling swan the yellow lore spot is much less evident at great distances and, as noted above, is rarely lacking altogether in whistling swans.

The rather high-pitched, whistling, barklike call, a one- to three-syllable *kow, kow-wow!* or *koe-wow!-wow*, with a fundamental frequency of about 1,000 Hz, and often uttered by flying tundra swans, is a useful guide for the field distinction of whistling swans, as compared with the lower-pitched and more powerful single- or double-noted and harmonic-rich calls of trumpeter swans. Wood, Brooks, and Sladen (2002) analyzed the vocal characteristics of trumpeter and tundra swans and their hybrids. Limpert and Earnst (1994) noted that the triumph ceremony (their "quivering wing" call) of tundra swans has three syllables lasting nearly two seconds, the middle one loudest, with up to five harmonics apparent on some sound spectrograms), which is often uttered when a pair reunites, or after an aggressive encounter, and is accompanied by rapid wing-quivering and sometimes ends with wing-flapping by one or both birds.

Bewick's swan

In the hand: Length 45–55 inches (115–140 cm). The Bewick's swan is the smallest of the northern swans; it is the only one with a partially yellow bill that has a wing measurement of under 575 millimeters in adults. The culmen length averages 91 mm (adult females) to 95 mm (adult males) (3.0–4.2 in) versus 101 mm (adult females) to 105 mm (adult males) in whistling swans (Rees, 2006). Females are identical to males, and juveniles have mottled gray and white plumage for most of their first fall and winter of life. Older immature birds gradually become indistinguishable from adults over the next year. Although the Bewick's swan always has a partially yellow bill, the yellow bill markings may reach but do not extend below and beyond the nostrils. If the yellow extends beyond the nostrils, and the culmen length exceeds 100 mm, the bird is likely to be a whooper swan. At least in England, many whooper swan juveniles have acquired an entirely white plumage by the time they are a year old, whereas in Bewick's swans some young birds retain traces of gray into their third year (Rees, 2006). *See the whooper swan account for further distinction criteria for Bewick's and whooper swans.*

Whistling swan–Bewick's swan distinction

Although, as indicated above, whistling swans average up to about 10 percent larger than Bewick's swans in all their usual bodily measurements, the two forms overlap to the degree that distinction between the them impractical using simple mensural criteria. In studies of birds from eastern North America and the Wildfowl Trust, the largest observed percentage of yellow in the lateral bill profiles of all whistling swans examined was 15.8 percent, or about one-sixth of the bill area, whereas the smallest proportion of yellow observed in Bewick's swan bill profiles was 22.9 percent, or almost one-fourth the bill area (Evans and Sladen, 1980; Rees, 2006). This fairly simple criterion is perhaps the only feasible way to distinguish the two races with a reasonable degree of confidence without resorting to a combination of several external measurements.

Evolutionary Relationships

The likely evolutionary relationships of the northern swans have been discussed earlier (Johnsgard, 1974). Although the current consensus is that the Bewick's and whistling swans are conspecific races of the tundra swan, there are some conflicting points that do not entirely support this merger, whereas in contrast to some earlier opinions (e.g., Delacour, 1954–64; Johnsgard, 1974) the trumpeter and whooper swans are now considered two distinct species.

Whooper × trumpeter swan hybrids have been produced in captivity, and a feral whooper swan has reportedly been observed to breed successfully with a feral whistling swan (McEneaney, 2004). Apparently wild intergrades between whistling and Bewick's swans have been seen in Japan, England, Saskatchewan, Oregon, and California, and mixed pairings among wild birds have been seen in various locations (Rees, 2006). Their behaviors associated with copulation are so similar that it is unlikely they would serve as behavioral isolating mechanisms. Degrees of fertility among these hybrid combinations are still unknown and would be of interest in judging relatedness in the group.

The Bewick's swan and whistling swan are currently (2019) classified as conspecific by most authorities, but this interpretation may be an oversimplification of the evolutionary history of this pair of northern swans. The similar small size of these two forms may simply reflect evolutionary convergence from separate stocks to comparable Arctic tundra environments, where selection for short fledging periods and associated small body masses might reflect convergent ecological adaptations rather than indicate conspecificity (Johnsgard, 1974). For example, Parkes (1958) and Livezey (1996) judged the Bewick's swan, rather than the whistling swan, to be the whooper swan's nearest relative. Given the fact that both intrahemispheric pairs of the northern swan taxa have a boreal forest breeder and a tundra breeder representative, it seems likely that speciation among them occurred fairly late during Pleistocene times, and that they are all closely related genetically. Their histories are perhaps somewhat comparable to the arctic-subarctic evolutionary histories of the lesser and greater white-fronted geese taxa (*Anser erythropus* and *A. albifrons*), the Ross's and snow geese taxa (*Anser rossii* and *A. caerulescens*), and the cackling and Canada goose taxa (*Branta hutchinsii* and *B. canadensis*).

Natural History

Breeding habitats, whistling swan

Whistling swans are associated with Arctic tundra throughout its breeding range in North America, and thus is an ecological counterpart of the Old World subspecies, the Bewick's swan. The whistling swan has a breeding range in Arctic tundra well to the north of the trumpeter swan's favored taiga breeding habitat, with very little breeding-season contact between them, a situation parallel to that of the Bewick's and whooper swans in Eurasia.

In Alaska, major whistling swan breeding areas include the Yukon-Kuskokwim delta (which probably supports about three-fourths of the subspecies' total population) and, to a much lesser extent, the Seward Peninsula and Kotzebue Sound area (Map 4). Some breeding occurs along the northern coast of Alaska, such as on the Colville River delta (150°E longitude). Whistling swans also breed on the north side of the Alaska Peninsula and adjoining Bristol Bay, on Kodiak Island, Nunivak and St. Lawrence Islands, and on Unimak and Izembek Islands in the eastern Aleutians (Baldassarre, 2014).

Those swans that breed south of the Brooks Range in western Alaska are part of the Pacific Coast population that winters mainly in western Washington, western Oregon, and the Central Valley of California (Rozell, 2017), although the relatively small population (ca. 4,000–5,000) of swans breeding on the Seward Peninsula and Kotzebue Sound is thought to include some that winter on the Atlantic coast. Those breeding north of the Brooks Range (from Point Hope and the Lisburne Peninsula northeast) all winter in the Atlantic flyway, collectively ranging from Maryland south to North Carolina, but primarily on Chesapeake Bay, along coastal Virginia, and on Currituck Sound, North Carolina. Milder winters recently are gradually allowing for more northerly wintering sites in Delaware or farther north.

The rest of the Alaska population (those breeding south of Point Hope) mostly winter in the Pacific flyway, chiefly in the Central Valley of California, although those breeding along Kotzebue Sound and the Seward Peninsula are apparently a mixture of birds using both flyway destinations (Map 4). During favorable winters swans also overwinter in interior regions such as the Great Salt Lake valley of Utah, where the yearly numbers there are influenced by the

Fig. 15. Whistling swan, adult landing

severity of the winters (Sherwood, 1960). Some weather-dependent wintering also occurs in southeastern Oregon and southern Idaho.

Limited wintering also occurs as far south as the lower Colorado River valley on the California-Arizona border. Rarely, whistling swans have also been reported in Mexico, with diminishing numbers reported by Drewien and Benning (1997) from Chihuahua, Tamaulipas, Durango, and Sonora. These authors also reported that band recoveries have been made from those same states, as well as from Baja California Norte.

The densest nesting concentrations in Canada are in the coastal Northwest Territories strip from the west side of the Mackenzie River delta to the east side of the Anderson River delta, with sparser populations inland, especially south of the tree line (Banko and Mackay, 1964). This population winters on the Atlantic coast (Sladen and Cochran, 1969), as do all the other Canadian breeding populations.

In central and eastern Canada, whistling swans are usually absent from the rocky landscapes of the Precambrian Shield in the Northwest Territories and Nunavut, but they do occur wherever typical lowland tundra occurs, north to Banks and Victoria Islands, the Boothia and Melville peninsulas, and south to about the Thelon River, Nunavut (64°N, 96°W). They also breed on southern Baffin Island and in Hudson Bay on Southampton, Coats, and Mansel Islands. In Manitoba they breed along the west coast of Hudson Bay south to Cape Churchill and in Quebec on the east coast of the bay south to about 60°N (Map 4).

A substantial number of whistling swans (600–1,000 birds as of the early 2000s) summer and breed on the Chukotka Peninsula of Siberia (see Map 4), where a whistling × Bewick's swan pair and a mixed pair with young have been seen (Rees, 2006). Small numbers of whistling swans have also been seen with Bewick's swans since 1990 at Japanese wintering areas, where several observations of apparent hybrids (Mikami, 1989) and mixed-pair birds have been documented (Rees, 2006).

Breeding habitats, Bewick's swan

The breeding habitat of the Bewick's swan consists of shallow tundra pools with abundant submerged vegetation and a luxuriant growth of shoreline vegetation (Voous, 1960). Rees (2006) stated that on the open tundra of Russia's northern coast nesting occurs on moss-lichen, moss-sedge, and sedge tundras and in low-lying marshes where pools and lakes are intersected by river channels. He also reported that on Vaygash Island, located about 60 miles north of mainland northeastern Russia, nesting occurs on shallow lakes that are surrounded by wide marsh plains with shrub tundra at higher elevations. There,

Whistling swan, adult

about one-fourth of the nests are built on relatively dry sites among willow bushes, whereas on the mainland open marine tundra tends to support higher swan densities than somewhat drier inland tundra, and there the swans generally nest in sedge meadows and moss-sedge habitats.

However, other regional variables in nesting densities have been found in Russian breeding areas that override any possible effects of such ecological variations in influencing overall nest densities. Among these variables are spring weather conditions, which has been shown to be a key factor in influencing the proportion of territorial pairs that build nests and attempt to breed. Additionally, cold weather during egg-laying and incubation may result in nest abandonments and thus lower nesting densities during otherwise early springs (Rees, 2006).

Winter Habitats and Foraging

Winter habitats, whistling swan

Normally the winter habitat of whistling swans includes sufficient aquatic plant life to provide adequate food but during unusually severe winter conditions feeding in cornfields has been observed (Nagel, 1965).

Preferred wintering habitat for whistling swans in the Chesapeake Bay area consists of open and extensive areas of brackish water no more than five feet deep (Stewart, 1962). January counts in that region during Stewart's studies indicated the following percentage usage of available aquatic habitats: brackish estuarine bays, 76 percent; salt estuarine bays, 9 percent; fresh estuarine bays, 8 percent; slightly brackish estuarine bays, 6 percent; and other habitats, 1 percent. Freshwater areas were used primarily by early fall arrivals.

Winter habitats, Bewick's swan

In Europe, Bewick's swans have traditionally wintered in areas with shallow coastal waters, coastal lagoons, inland freshwater lakes and marshes, and flooded pastures. In these places they forage mostly on submerged macrophytic plants, especially the rhizomes and tubers of pondweeds (*Potomogeton* spp.), horned pondweeds (*Zannichellia* spp.), and stoneworts (*Chara* spp.). They also eat the roots and rhizomes of eelgrass (*Zostera* spp.) in tidal basins and the roots and upperparts of grasses and forbs when foraging on wet or dry pastures (Rees, 2006).

With habitat changes associated with European wetland reclamation, drainage, and water pollution during the latter half of the twentieth century, the swans increasingly moved to foraging on arable lands and pastures, and gradually increased their use of stubble for fall foraging. They also especially began

Whistling swan, adult

using winter cereal crops, root crops, and oilseed rape (*Brassica napus*) during winter months. In a British study, overall October to March use was greatest on coastal waters, followed by permanent inland waters, improved pastures, and arable lands (Rees, Kriby, and Gilburn, 1997). However, during winter the birds tended to concentrate in larger groups on relatively few arable sites (flock sizes averaged 90.5 birds in a January study), but also dispersed widely to forage on inland waters (average flock size 41.6 birds) and on improved pastures (average flock size 38.5 birds) (Rees, 2006).

Where whooper swans also occur during winter they might locally displace the smaller Bewick's swans from favored foraging sites. On their breeding grounds whooper swans typically occupy taiga habitats well to the south of the tundra habitats used by Bewick's swans, but recently whooper swans have moved north-ward into tundra habitats previously used by Bewick's swans, perhaps as a result of climate change, and such movement may in the future be to the detriment of Bewick's swans (Rees, 2006). A similar possibility of future expanded contact exists for trumpeter and whistling swans in north-central Alaska.

Winter foods and foraging, whistling swan

Like other swans, the whistling swan feeds predominantly on vegetable ma-terials from aquatic plants. Foods taken by whistling swans on the breeding grounds are not yet well studied, but in migration and wintering areas the birds usually feed extensively on the roots, stems, and leaves of such aquatic plants as wild celery (*Vallisineria* spp.), wigeon grass (*Ruppia* spp.), bulrushes (*Typha* spp.), and pondweeds (*Potomogeton* spp.). The tubers of arrowhead (*Sagittaria* spp.) are also favored foods, and in brackish waters the birds may feed to some extent on mollusks, especially clams.

Martin, Zim, and Nelson (2011) listed grasses and sago pondweed (*Potamogeton pectinatus*) as major food for both the Eastern and Western Populations of whistling swans, and additionally listed wild celery (*Vallisineria americana*), lady's thumb (*Polygonum persicaria*), horsetail (*Equisetum* spp.), and bur reed (*Sparganium* spp.) as important foods in one region or the other. Sherwood (1960) reported that tubers and seeds of sago pondweed had been the exclu-sive food of 12 whistling swan specimens obtained in Great Salt Lake valley, Utah, although other aquatic foods were locally available.

Stewart and Manning (1958) and Stewart (1962) reported on the winter foods of whistling swans in Chesapeake Bay and found that birds foraging in the pre-ferred brackish estuarine bay habitat relied largely on wigeon grass (*Ruppia*) and to a lesser extent on sago pondweeds, with bivalve mollusks (*Mya* and *Macoma*) also being taken in considerable amounts by birds foraging in marine habitats. Among a sample of 49 birds, submerged vegetation was the only food consumed in freshwater habitat, as compared with 60 percent in brackish water and 41 per-cent in estuarine ponds. Four birds collected in freshwater estuaries had been feeding almost exclusively on wild celery, while four obtained from estuarine marsh ponds had been eating wigeon grass, three-square (*Scirpus*), and grasses.

Martin, Zim, and Nelson (2011) and others have suggested that whistling swans may despoil the supply of duck foods in some areas, and certainly the preferred foods such as sago pondweed and wigeon grass are also used by many ducks. Wigeons and canvasbacks are species with habitat preferences and foods similar to those of whistling swans in the Chesapeake Bay region (Stewart, 1962). Sherwood (1960) mentioned observing a considerable num-ber of species of geese and ducks feeding among whistling swans without any visible intolerance on the swans' part. He suggested that the swans may actually increase the forage for the ducks, both by pulling up more food than they actually consume and by possibly creating new sago beds through the dissemination of seeds and tubers as well as by "cultivation" of the marsh bottom.

Thompson and Lyons (1964) noted that pronounced daytime foraging flights were not characteristic of the spring flock of whistling swans they studied, reporting that average midday counts were only about 200 birds fewer than average morning and evening counts (749 and 771 birds, respectively). Sladen and Cochran (1969) observed that wintering whistling swans rarely reached an altitude of 1,000 feet during local movements to and from foraging areas. Migrating swans often reach much higher elevations; whooper swans have been documented at an elevation of 8,000 meters (26,250 feet) (Stewart, 1978).

Field feeding by whistling swans in Utah was first reported by Nagel (1965), who observed them feeding on corn during periods when ponds were iced over. This behavior has also been noted on Maryland's eastern shore, where feeding in cornfields on waste kernels was reported by Munro (1981a). They have also been found to locally feed on soybeans and on the seed heads of winter wheat.

Winter foods and foraging, Bewick's swan

A substantial amount of grazing on agricultural lands is currently performed by wintering Bewick's swans in Britain, where they consume grain, waste potatoes, and sugar beets. In England, various pasture grasses (*Glyceria, Agrostris*, etc.) are grazed extensively when they are wet or flooded, and some grazing of grasses of drier pastures is also done (Owen and Kear, in Scott and the Wildfowl Trust, 1972; Rees, 2006).

During my research in the early 1960s at England's Wildfowl Trust (now the Wildfowl and Wetlands Trust), wintering Bewick's swans typically roosted on the mudflats of the nearby River Severn and flew in twice daily to the Trust grounds to eat grain that had been put out for them. Sometimes they would stay at the Trust all day, returning to the river only after the late afternoon feeding period. During the half century that has passed since then, Bewick's swans have increasingly relied on agricultural lands for their winter foods, as wetland habitats have disappeared and agricultural produce, aided by the use of fertilizers and other advances, has become more widely available and perhaps also more nutritious.

Little specific information on breeding-ground foods is available, but in Russia the birds are reported to consume a variety of aquatic plants and grassy territorial plants. As with the other northern swans, the leaves, roots, and stems of pondweeds are favored foods on migration and in wintering areas, and the roots and rhizomes of eelgrass (*Zostera*) are also an important food.

Whistling swan, adult head profile

Social Behavior

Age of initial pairing

Very little reliable information is available on the age of sexual maturity in whistling swans and Bewick's swans; they have been bred only rarely in captivity. Delacour (1954–64) reported breeding by a five-year-old female with an older male, and Robert Elgas (pers. comm.) successfully bred a pair of hand-reared whistling swans when they were six years old. Two pairs of swans hatched from wild-taken whistling swan eggs initially nested when they were four years old (William Carrick, pers. comm.).

In the Bewick's swan, initial pairing occasionally occurs when the birds are two years old (their third summer) but more often occurs at three or four years of age, with females usually pairing slightly earlier than males. Bewick's swans thus usually pair (and initially breed) a year later than do whooper swans and mute swans (Rees, 2006). Rees also reported that of 123 Bewick's swans studied over several winters and known to have paired, 34 percent were associated with mates at age two, 69 percent by age three, and 91 percent by age four. Initial pairing was seen in birds as old as nine years. The mean age of pairing was 3.18 years for females and 3.5 years for males.

The triumph ceremonies (also called "greeting ceremonies" and described in detail for the whooper swan by Brazil, 2003) and behavior associated with pair-bonding and copulation are nearly the same in tundra swans as in those typical of trumpeter and whooper swans, although the speed of movements and associated vocalizations among them differ appreciably (Johnsgard, 1965). Impediments to intertaxon pairing would seem to be dependent on these relatively minor behavioral differences, although the pair-bonding process in swans is evidently so prolonged that any incipient pairing "mistakes" would probably be corrected in time, assuming the presence of available birds of the same taxonomic group.

Spring flock composition

Flock sizes in tundra swan wintering areas and during migration are often large and may number in the hundreds or even the thousands, although the birds become strongly territorial and well scattered during the breeding season. The birds have relatively long migratory routes, often more than 2,000 miles between wintering and breeding grounds.

Counts made during spring in Wisconsin indicate that most whistling swan flocks consist of units of up to as many as about 13 birds that remain together on local foraging flights. Thus, multigenerational families and pairs are the obvious unit of substructure in whistling swan flocks. By the time they reach their breeding grounds these flocks have broken up and the birds spread out widely over the tundra, often in densities of only about one or two pairs per square mile.

Reproductive Biology

Pair formation and pair-bonds

As for the other swans, the pair-bonds of this species appear to be strong and potentially permanent. Peter Scott (1972) reported no observed cases of "divorce" among hundreds of individually recognizable Bewick's swans over seven years of observation and that up to three years had been required for bereaved swans to take a new mate.

Whistling swan, adult incubating

Dafila Scott (1967) reported that some Bewick's swans left in the spring with one mate and returned the next fall with a different one, suggesting that mate replacement sometimes occurs during a single breeding season. Some tentative pairing may occur during the second winter, but in six of seven cases that she observed, these pairings had broken up by the following winter. Peter Scott (1972) noted, however, that some swans may remain with their parents for their second or even third winter of life, and one female was observed still associating with her parents at Slimbridge when she was six years old (Rees, 2006).

Through this work it is known that pair-bonds are strong and permanent, without a single known case of "divorce" among hundreds of pairs studied. Birds that lose a mate may take up to three seasons to find a new one, but some establish new bonds much sooner. Further, family bonds persist in immature birds even up to the third winter of their lives, resulting in the association of up to four generations of related birds.

Some tentative pairing may occur in birds as early as their second winter of life, but nearly all these bonds are broken by the following winter, and probably initial breeding does not occur in Bewick's swans until the birds are nearly four years old (Scott and the Wildfowl Trust, 1972). Peter Scott (1966) noted that two-year-old Bewick's swans spent quite a lot of time in courtship display during the winter months. However, Dafila Scott (1967) mentioned that many of the pair-bonds formed during second-winter birds were only temporary and usually were broken by the following winter. Pair-bonds are usually established when the birds are two or three years old, during the bird's third summer, with females typically pairing earlier than males. Pairing between siblings has not been reported. Pairing and breeding usually occur a year earlier than is true of whooper and trumpeter swans, a trait that has been related to the stresses associated with the Bewick's swan's long migration and its short available breeding season in the high Arctic (Rees, 2006).

Pair-bond durations were also studied among Bewick's swans at the Wildfowl Trust's Slimbridge and Welney observation sites. The durations varied over periods ranging from 1 to 19 years and averaged 3.47 years for 331 pairs in which the actual duration was known. For nearly 39 percent the duration lasted only a single year, perhaps in part because of the relatively high annual mortality rate of 15 percent for this population—there is a roughly 30 percent chance of one of the pair members dying during the course of a year. However, 34 percent of the paired birds remained together for at least five years, an 8 percent rate for at least ten years. One pair was together for 20 years.

These long pair-bonds are obviously contingent on extended lifespans. The average lifespan of wild British Bewick's swans has been calculated at 5.4 years (Rees, 2006), but a maximum longevity of 29 years has been reported. Annual overall survival rates for this population have been estimated at 87.1 percent and 82.2 percent, with first-year juveniles having lower rates (64 percent for males vs. 68 percent for females) as compared with survival rates of more than 80 percent for older age categories (Scott, 1988; Rees, 2006). These survival rates are fairly close to those reported for mute swans in Britain (McCleery et al., 2002).

Annual survival rates of the whistling swan, based on 5,963 neck-banded individuals captured in Maryland and North Carolina, were estimated to be 92 percent for adults (both sexes), 81 percent for juvenile males, and 52 percent for juvenile females (Nichols et al., 1992). Bart et al. (1991) estimated a 52 percent survival rate for juveniles during their first fall migration, and a 76 percent subsequent survival rate over their first winter.

Only three cases of "divorce" among Bewick's swans were recorded during these studies, but because of their long lifespans and substantial mortality rates, many individual swans have multiple mates over their lifetimes. Of 3,280 birds, 91.2 percent had a single mate, 6.3 percent had two, 1.2 percent had three, and 0.4 percent had four or more mates, with a maximum of seven mates recorded. On average, 2.6 years elapsed between the time a paired bird's mate was absent and when it returned to the wintering grounds with a new mate, as compared with 1.9 years for Icelandic-breeding whooper swans. Evidently Bewick's swans change mates less frequently than either mute swans or whooper swans; among birds observed for as long as five years, 15.8 percent of mute swans, 7 percent of whooper swans, and 3.5 percent of Bewick's swans were observed with two or more mates (Rees et al., 1996; Rees, 2006).

As in all other swans, pairs are formed and maintained by mutual displays such as the triumph ceremony (Fig. 7); copulation probably plays little if any role in such pair-bond development, as it rarely if ever occurs among wintering flocks. Pair formation is a gradual and inconspicuous process, with a major aspect being the tendency of males to defend mates or potential mates and, after expelling intruders, to return to the female, where they join in a mutual triumph ceremony (Johnsgard, 1965). Differences in the head shape and bill patterning are apparently important bases for individual recognition among the Arctic-breeding swans, and it is highly probable that individual differences in vocalizations may also play a role in mate recognition.

Although winter courting activities occur in wintering birds, actual pair formation and copulation have not been reported among wintering Bewick's swans in Britain. Evidently some if not most of pair-forming activities occur within the breeding range among nonbreeding flocks, especially birds in their third or fourth year. Copulations have been observed in spring migratory staging areas and might occur several times daily after breeding birds are settled on their nesting territories. Nonbreeders that have established territories but did not then build nests have also been seen copulating in early to mid-June. Among breeding pairs, the eggs are laid in late May to early June (Rees, 2006).

Like the trumpeter swan, mutual head-dipping movements that closely resemble those of bathing birds precede copulation in tundra swans. Unlike the mute swan, preening movements do not play a role in precopulatory behavior. As treading is terminated, the male releases his grip on the female's nape as both birds extend their necks strongly upward and utter loud notes, usually simultaneously extending and waving their wings in the same manner as during triumph ceremonies. (Johnsgard, 1965).

Whistling swan, juvenile head profile

Nesting onset

Among North American whistling swans, arrival on their tundra nesting grounds of western Alaska after leaving their California wintering grounds occurs in late May, and nesting is usually underway by the first of June, as the ponds, creeks, and other wetlands become ice-free. There is a high degree of synchrony of nest initiation and egg-laying in individual areas, so that most incubation begins (and hatching occurs) over a period of only three or four days. Most pairs choose nest sites on the shore of a lake or a pond within 20 yards of water; somewhat fewer nests are built on islands or points of land in lakes, and even fewer on tundra or in other locations. Elevated sites, such as hummocks, are favorite nest sites and are also among the first areas to be snow-free in spring.

Bewick's swans likewise winter at great distances, usually well over a thousand miles, from their breeding grounds, and their departure from England for the breeding grounds occurs during January and February. Arrival on the breeding grounds of Siberia may not occur until May and is associated with the onset of thaws and the appearance of flowing water in rivers (Dementiev and Gladkov, 1967). The Bewick's swans wintering in Britain typically arrive on their Russian breeding grounds in early to mid-May. Experienced breeding pairs begin their prenesting activities while their territories are still largely covered by ice and snow, whereas first-time breeders are likely to wait until thawing has occurred (Rees, 2006).

Territorial defense

Although an average of about 73 percent of the Bewick's swans wintering in Britain are paired, not all of the paired birds arriving on the breeding grounds succeed in, or perhaps even attempt, to establish breeding territories. Even for those pairs that do achieve in establishing a territory, not all succeed in building nests. A good deal of time and energy is spent in expelling intruding swans, including aerial chases and outright fighting. Losers in such encounters are likely to abandon their territories and join the flocks of nonbreeding birds. Additionally, adverse weather during the early nesting season might cause many pairs to abandon their breeding efforts. During one three-year study of Siberian coastal tundra, the percentages of territorial pairs with nests in June varied from 20 to 71 percent, and the percentages of those still attempting to breed in July ranged from 20 to 74 percent. These figures correspond fairly well with the percentage of pairs observed arriving annually with cygnets (12–66 percent) (Rees, 2006).

Nesting densities

Lensink (1973) reported nesting whistling swan densities of from 130 to 320 hectares per pair (0.8–2.0 pairs per square mile) in the Yukon–Kuskokwim delta of western Alaska. Swan densities based on aerial surveys in Canada's Northwest Territories were estimated from 1948 through 1953. In the Mackenzie River delta, densities averaged 1.5 swans per square mile. In the area between the Mackenzie and Anderson Rivers, the comparable averages were coastal tundra, 1.7; upland tundra, 1.3; and transition zone to coniferous forest, 0.3 (US Fish and Wildlife Service, 1954).

In 1950 the area from the Annak River to Kent Peninsula of Nunavut was found to have a swan density of 0.16 per square mile, whereas southwestern and southeastern Victoria Island had a much lower swan density of only 0.007 per square mile. It would seem that a maximum density of about one or two pairs of whistling swans per square mile might be expected in very favorable lowland tundra habitats.

Several estimates have been made for nesting densities across the Russian Arctic and summarized by Rees (2006). In four such studies, the estimated densities of nests per square kilometer were 0.18, 0.20, 0.22, and 1.10 (0.47, 0.52, 0.57, and 2.85 nests per square mile); the last of these surveys was done in a nature reserve. Spring weather has a strong influence on nest densities, with cold weather during egg-laying and incubation having a marked negative effect.

Whistling swans, adults in flight

The nests of whistling swans are typically well scattered over the tundra. Banko and Mackay (1964) reported that nest sites vary in location from the edge of water to the top of low hills a half-mile from water, with small islands in tundra ponds being preferred locations. Nests are usually mounds of moss, grasses, or sedges and are from one to two feet high (Banko and Mackay, 1964).

The nests of Bewick's swans are likewise constructed of tundra vegetation, especially sedges and mosses, with a lining of down. It is typical that the pair uses an old nest site after some refurbishing, with the female lining the nest with down or other feathers (Dementiev and Gladkov, 1967).

Nest-site fidelity

The only data on nest-site fidelity comes from a five-year Russian study by Shchadilov et al. (1998). Among 184 pairs whose members were both identified, two (1.11 percent) were ones in which both members returned to the same nest site for all five years, eight (4.3 percent) returned at least four years, 22 (12 percent) returned at least three years, and 39 (21.2 percent) returned for at least two years. Pairs usually retired to the same territory they had defended the previous year: among 111 pairs, of which at least one of the members had been seen there previously, 92 percent returned to the same territory.

Egg laying and clutch sizes

Egg laying begins shortly after arrival at the whistling swan's tundra breeding grounds in late May or early June (Banko and Mackay, 1964). In southeastern Victoria Island, at the northern edge of this subspecies' range, the nests are constructed in as little as five days, and in one case a nest was built and three eggs were deposited in no more than eight days (Parmelee, Stephens, and Schmidt, 1967). Egg-laying in Alaska begins in late May or early June, and likewise the first eggs of Bewick's swans are laid on Novaya Zemlya at the end of May or early June.

The average clutch size of 297 whistling swan clutches was 4.3, with a mode of 5 and a range of 1 to 7. According to Banko and Mackay (1964), 4 eggs constitute the whistling swan's normal clutch, with as many as 7 eggs present at times. Such large clutches are a possible result of laying by two females. There are marked year-to-year differences in average clutch sizes of both whistling swans and Bewick's swans, with smaller average clutch sizes associated with late, cold springs. Lensink (1973, and in Scott and the Wildfowl Trust, 1972) reported that 5 eggs were the normal clutch size in good springs, but only 3 or 4 eggs were usually laid in cold, wet springs. Fischer and Stehn (2015) reported a ten-year average of 4.3 eggs. Clutch sizes of 427 nests found in the Yukon-Kuskokwim delta during the 1963–79 period averaged 4.3 eggs, with the lowest average being 3.3 eggs in 1971 and highest 5.2 eggs in 1978 (Dau, 1981; Pacific Flyway Council, 2017).

Bewick's swan, adult

Clutch sizes of high-latitude Bewick's swans on Novaya Zemlya (ca. 75°N latitude) have been reportedly only 2 to 3 eggs (Dementiev and Gladkov, 1967), and studies on Vaygash Island (70°N) also indicate annual clutch-size averages ranging from 2.3 to 3.3 eggs (Rees, 2006). Both of these sets of data suggest that clutch size in this high-Arctic breeder might be influenced by latitude since studies from Russia generally have indicated a mean clutch size of 3.62 eggs, with a range of 1 to 6, the year-to-year variations depending on climate and food conditions (Kear, 2005). Typically there is a 48-hour interval between the laying of each egg, although the last egg of a clutch might have an interval of up to 72 hours before being deposited. In a 5-egg clutch, up to eight days might pass between the laying of the first egg and the onset of incubation (Rees, 2006).

Incubation and hatching

Among North American whistling swans, an incubation period of 30 to 32 days is typical; Banko and Mackay (1964) likewise estimated the tundra swan's average incubation to last about 32 days. Robert Elgas (pers. comm.) noted a 30-day incubation period for Alaskan tundra swan eggs incubated under geese. Hatching in Alaska and Canada begins in late June or early July, and young are probably still present into September.

In Bewick's swans an incubation period of 29 to 30 days is usual, with eggs typically hatching over a 24- to 72-hour timespan, a duration that is longer than that of most waterfowl. Sometimes the female will carry the shell of the first-hatched egg over a distance of more than 100 yards and drop it, a behavior that has rarely, if ever, been documented in other waterfowl. Hatchling success is typically very high in wild Bewick's swans, with estimates of 89.8 and 91.8 percent of the eggs hatching in two Russian studies (Rees, 2006). However, overall nesting success (the percent of incubated eggs that hatch successfully) varied over a six-year study period on Vaygash Island from 20 to 94 percent, with the success rate lower in years that had cold weather during laying or late springs or both.

Brood rearing and fledging success

The best information on brood sizes and losses comes from Russian studies summarized by Rees (2006). In a sample of 495 Bewick's swan clutches from Europe and northwestern Russia there was an average clutch size of 3.1 eggs, as compared with an average brood size of 2.7 cygnets. The breeding experience of the parents might be an important factor in influencing the survival of newly hatched broods through their varied effectiveness in protecting them from predators such as arctic foxes (*Alopex lagopus*). Additionally, relative previous parental breeding experience (the number of years a pair has been together) seems to be important in influencing the number of Bewick's swan cygnets that are not only fledged but also are brought safely to wintering grounds—although this positive parental influence on cygnet survival was not apparent in some years.

No doubt a critical relationship exists between the time of fledging and the first freezing weather, which may greatly influence fledging success of whistling swans during some years. Most broods remain within 100 to 400 yards of the nest for some time, and the young grow at a rapid rate, reaching weights of 11 to 12 pounds in 70 days. They fledge at about the same time, at 60 to 75 days of age. Banko and Mackay (1964) reported that hatching occurs in late June or early July, and fledging occurs about the middle of September.

Although a fledging period of only 30 to 45 days was once reported for Bewick's swans (Dementiev and Gladkov, 1967), more recent studies indicate that they fledge in 60 to 70 days, which is comparable to the estimated 60- to 75-day fledging period of whistling swans. Nonetheless, since the breeding territories are abandoned in September, Bewick's and whistling swan cygnets can be little more than 80 to 90 days old before they must begin their thousand-mile-plus southward migration.

Average brood sizes typically range from 2.55 to 2.63 young per pair in Alaska and the Northwest Territories. Banko and Mackay (1964) estimated that an average of only 2 or 3 whistling swan cygnets survived from hatching until fledging in autumn.

Post-fledging juvenile and adult mortality

By counting the percentage of the distinctively plumaged juveniles during fall and winter, estimates of annual productivity and mortality can be attained. Winter brood counts of whistling swans have generally ranged from 2.15 to 2.63 cygnets per pair, suggesting a cygnet hatching-to-winter-mortality rate of about 18 to 25 percent. During the eight-year period between 1964 and 1971, the average number of cygnets per winter family varied from 1.54 to 2.24 (avg. 1.93), and the percentage of juveniles of the Atlantic coast wintering population ranged from 4.8 to 14.6 percent (avg. 11.1) (Lynch, 1972). Chamberlain (1966) noted that the percentage of first-year whistling swans in the 1964–65 winter flocks on Chesapeake Bay ranged from 9.46 to 13.9 percent, while in 1965–66 the percentages were slightly lower and ranged from 8.22 to 12.1 percent.

Estimated percentages of juveniles in the major Atlantic Coast wintering states (New Jersey, Maryland, Virginia, and North Carolina) from 1961 through 2005 were 1961–69, 11.6 percent; 1970–79, 13.0 percent; 1980–89, 16.5 percent; 1990–99, 11.3 percent; and 2000–05, 9.8 percent (Caswell et al., 2007), for an overall unweighted 1961–2005 average of 12.4 percent. In 2017 juveniles composed 12 percent of the whistling swan's Eastern Population, as compared

Bewick's swan, adult head profile

with a long-term average of 13 percent (US Fish and Wildlife Service, 2018). Assuming a stable population, these data would suggest an immature and adult annual mortality rate of about 9 to 14 percent. The percentage of young was highest during January counts because of the typically later arrival to wintering areas of family groups than of nonbreeders.

Fall counts in Utah of whistling swan age groups from 1963 to 2011 averaged 2.2 juveniles per family, suggesting an approximate 50 percent summer-to-fall loss relative to typical clutch size (Pacific Flyway Council, 2017). Similarly, Bart et al. (1991) estimated a 48 percent loss of young during their fall migration, and a 24 percent mortality thereafter during their first winter. Over the period from 1962 through 2015, the proportion of juveniles in fall flocks of the Pacific Flyway population has declined from about 40 percent to about 20 percent. It was suggested that density-dependence factors that affect rates of territorial occupancy, productivity, or juvenile survival might account for this declining trend (Pacific Flyway Council, 2017).

Information on adult mortality rates in whistling swans is sparse because relatively few adults are banded and they have been legal game in only a few states. Based on sightings of nearly 6,000 neck-banded birds, survival in the Eastern Population was judged to be 92 percent for adult males and females, 81 percent for juvenile males, and 52 percent for juvenile females (Nichols et al., 1992). Wild female whistling swans have been known to live for as long as 23 years (Baldassarre, 2014).

As of the early 2000s, the number of hunting permits issued for the Eastern Population states (and three western states where hunting was legal) totaled North Carolina, 5,000; North Dakota, 2,000; South Dakota, 1,500; Virginia, 600; and Montana, 500. The average number estimated to have been killed during the 2003–5 hunting seasons were North Carolina, 2,179; North Dakota, 575; Virginia, 178; South Dakota, 104; and Montana, 85. The average overall total kill for the Eastern Population from 1983 to 2005 was 3,600 birds, and a similar annual kill rate of 3,313 birds was estimated for the three-year period 2003–5 (Caswell et al., 2007). Assuming an Eastern Population of about 100,000 birds during the early 2000s, sport hunting was likely to be responsible for about 3 to 3.5 percent of the species' total first-winter and older Eastern Population birds, and about 25 percent of the population's overall mortality, assuming an adult annual mortality rate of about 9 to 14 percent.

Much information on the Bewick's swan relative to its survival and longevity in the wild has been obtained from the returns of individually recognized birds to the Wildfowl Trust in later years. Evans (1977) provided an analysis of such sightings for a seven-year period for birds that were adults or second-year birds when first sighted and could be recognized individually. Of a total of 792 identifiable birds, 287 were seen the subsequent winter season, indicating a minimum annual survival rate of 36.2 percent for adults and subadults, and a probable much higher actual rate, given the likelihood of some overlooked birds or birds that wintered outside the survey area. More recent studies of birds returning in their third and subsequent winter seasons suggest an annual survival rate of 87 percent (Kear, 2005; Rees, 2006).

Remarkably, 33 percent of these older age classes of Bewick's swans were found to return seven years after initially having been first identified. The astonishing number of birds at least nine years old reported in this study indicates that the survival rate of these swans is quite high, in spite of local hunting and other mortality threats occurring during their long migrations. Wild Bewick's swans have been documented to live as long as 29 years (Rees, 2006),

Bewick's swans, adult with downy cygnets

a probable longevity record for wild individuals of any of the northern swans, and a record made more remarkable considering this species' remarkably long migration routes.

Observations on wintering birds at the Wildfowl and Wetlands Trust indicate that Bewick's swan offspring will rejoin their parents, might overwinter with them, and depart with them in the spring up to at least their fifth winter of life (Kear, 2005), which is an indication of the likely importance of social bonding, and probably also evidence of the values in such prolonged intergenerational contacts in long-term maintenance and transmission of important survival information.

Molting, migration, and wintering

In Alaska, adult whistling swans undergo their flightless period in July and early August, while nonbreeders molt earlier and begin to regain their power of flight in late August. The postnuptial molt of breeding adults occurs while the young are still flightless, the female becoming flightless within about two weeks after the young hatch, while the male molts somewhat later, often about the time the female has regained her flight (Banko and Mackay, 1964; Wilmore, 1974). Assuming each might be flightless for about five weeks, the adults should both have regained their power of flight by the time the young are about 80 days old and have also fledged.

By about 85 days after the peak of hatching, families with fledged young begin to join nonbreeders for the fall flight southward (Bellrose, 1976). At that time, about mid-September, a fairly leisurely fall migration southward begins through the interior of northwestern Canada along the Mackenzie River valley. By early October large concentrations of birds occur on Lake Clair and Richardson Lake in northeastern Alberta.

Ely et al. (1997) established the migration routes of Western Population whistling swans breeding on the Yukon-Kuskokwim delta. One of the routes follows an interior migratory corridor. Leaving in late September, the birds cross the Alaska Range and stop briefly at the Susitna Flats of the Upper Cook Inlet. They then fly into Yukon, north of and parallel with the Wrangell Mountains, to a staging area in northeastern British Columbia. From there they pass through central Alberta, southwestern Saskatchewan, and Montana to a staging area in southeastern Idaho. In early December they migrate to the Sacramento–San Joaquin delta of California. The return spring flight follows the same 2,000-plus-mile corridor.

For those Alaskan birds that migrate on the interior route, there is some contact between swans from the Western Population and Eastern Population in southwestern Saskatchewan. There, western population birds turn west, and Eastern Population swans turn east (Ely et al., 1997).

Some migrants of the Western Population, probably mostly from Bristol Bay and the Alaska Peninsula, pass eastward through south-central Alaska, sometimes stopping briefly at Cook Inlet and on the Copper River delta, then enter southeastern Alaska along coastal or interior routes. Most then continue on southeast across British Columbia to Alberta, joining other Alaskan birds also following interior routes, plus some Eastern Population swans from the Mackenzie River drainage and Alaska's North Slope. The remaining swans follow a coastal route from southeastern Alaska through British Columbia and south to central California (Pacific Flyway Council, 2017).

Those Western Population swans that pass through southern Alberta continue on along one of two southward routes. Some continue south by early November, with up to 5,000 swans staging briefly at Freezeout Lake in north-central Montana. They then move on to northern Utah by mid-November and to northwestern Nevada by middle to late December. Another more western route takes birds from southern Alberta through Idaho, Malheur Lake, the Willamette Valley, and the Klamath basin of Oregon to the deltas of the San Joaquin and Sacramento Rivers. In recent years they have expanded into the American River basin of the lower Sacramento valley to forage on flooded rice (Pacific Flyway Council, 2017).

A few hundred swans that breed on the Alaska Peninsula winter on Unimak Island and Izembek Lagoon, but many winter in southeast Alaska. About 300 to 500 winter along coastal British Columbia, about 5,000 winter in coastal Washington (mostly on the Skagit River flats), about 10,000 winter in coastal Oregon (especially along the Columbia River), and about 5,000 winter on Great Salt Lake. Collectively, excluding Alaska and British Columbia, about 82 percent of the Western Population winters in California, followed by Oregon (7 percent), Utah (7 percent), Washington (3 percent), and Nevada (few). Sometimes Idaho, Montana, Wyoming, and Arizona also support some wintering swans, depending on water conditions (Pacific Flyway Council, 2017).

Birds breeding along Canada's Arctic islands as far as Banks Island fly south-southeast to join the Eastern Population birds from western Canada and Alaska that are headed for the Atlantic coast as well as birds arriving from the west coast of Hudson Bay. Baffin Island breeders, those from eastern Hudson Bay, and probably also the Southampton Island birds migrate almost directly south along the eastern shore of Hudson Bay on through the Great Lakes region to Atlantic coast wintering grounds, and possibly also to wintering grounds along the south shore of Lake Erie and in eastern Pennsylvania, less than 2,000 miles away from Baffin Island.

The whistling swan flocks that migrate from western and central breeding grounds in Arctic Canada gather successively at a variety of fall staging areas across the northern Great Plains (Wilkins et al., 2010), including northern and southern Alberta, southwestern Saskatchewan, the Devils Lake region of North Dakota, and the Upper Mississippi River National Wildlife and Fish Refuge in southeastern Minnesota (see large arrowheads on Map 4). Leaving the upper Mississippi River, they make nonstop 1,000-mile flights to Chesapeake Bay and coastal North Carolina, completing an overall migration route that began 3,000 to 3,500 miles away on Alaska's Yukon-Kuskokwim delta. The highest US Audubon Christmas Bird Count tallies in recent years have been at Pettigrew State Park and Mattamuskeet National Wildlife Refuge in North Carolina, where the count numbers have ranged from about from about 16,000 to 37,000 birds.

The Bewick's swans wintering at the Wildfowl Trust in Britain make their 3,000- to 3,500-kilometer (1,860- to 2,175-mile) flights to probable breeding grounds on Kolguyev Island in the Barents Sea with only two or three stop-over rests. They can cover some 1,450 kilometers (900 miles) before needing to stop, rest, and forage (Rees, 2006).

Status

Whistling swan status

The North American whistling swan population probably consisted of somewhat less than 100,000 birds in the 1970s, judging from federal surveys. Waterfowl breeding surveys in Alaska began in 1955, and the six-decade (1955–2015) mean total annual swans recorded was 62,000 swans. Slightly over half of the total continental population nests in Alaska. Alaska is also the only region where whistling swans that breed south of the Brooks Range and winter in the western states and provinces (the Western Population), and those swans that breed almost entirely north of the Brooks Range and winter in eastern North America (the Eastern Population), come into limited breeding season contact. This restricted breeding region of swans having mixed migratory traditions occurs in the Seward Peninsula–Kotzebue Sound region, and probably involves less than less than 3 percent of the total Alaska breeding population (Ely et al., 1997; Pacific Flyway Council, 2017). The breeding areas of Alaska's Western Population and their 1970s estimated breeding populations included Yukon-Kuskokwim delta (53,000), Bristol Bay (12,200), Kodiak Island (100), St. Lawrence Island (100), Izembek–Unimak Island (60), and Nunivak Island (50), totaling about 65,000 breeders.

The only Alaskan breeding region having entirely Eastern Population migrants is the Arctic coastal tundra region north of Brooks Range, which supports about 3,000 to 4,000 birds (Pacific Flyway Council, 1989). Breeding areas that have a mixture of Western and Eastern Population birds, with up to about 8 percent of the birds migrating to Eastern Population wintering grounds, include the Seward Peninsula with about 2,800 birds and Kotzebue Sound with about 1,800 birds (Pacific Flyway Council, 2017). Based on a recent ten-year average, it was estimated that about 3 percent of all the whistling swans that breed within the Western Population's breeding region (south of Alaska's Brooks Range) migrate to Eastern Population wintering grounds (Pacific Flyway Council, 2017).

As of 2015, the ten-year average for breeding surveys in Alaska was 123,426 birds, with an all-time high of 174,428 birds in 2008 (Pacific Flyway Council, 2017). Nest surveys done on the Yukon-Kuskokwim delta from 1985 to 2014

Bewick's swans, adult with downy cygnets

also indicate an increasing trend in swan nesting numbers; between 2004 and 2014 the estimated number of nests found in the coastal zone averaged 12,152, with the highest number being 18,987 in 2014 (Fischer and Stehn, 2015).

In order of their relative breeding importance, the three-year 2014–16 estimates of population averages (their ranges are shown in parentheses) of breeding-season whistling swans for five major Alaskan nesting regions are: Inland Yukon-Kuskokwim delta, 48,969 (31,783–60,206); coastal Yukon-Kuskokwim delta, 27,221 (23,000–31,351); Kuskokwim Sound, 14,713 (8,491–24,744); Bristol Bay, 13,200 (11,181–14,635); and Seward Peninsula, 7,685 (6,875–8,800). The collective 2014–16 average for all five areas was 111,535 (89,177–129,102) swans

(Pacific Flyway Council, 2017). In 2018 the total swan index for the Western Population was 152,100 birds, with no statistically significant trend evident during the previous ten years (US Fish and Wildlife Service, 2018).

The majority of the Eastern Population's breeding areas are in Arctic Canada. Detailed quantitative estimates of Canada's many geographic components of Eastern Population whistling swans are not available, but its overall breeding expanse extends nearly 2,000 miles across Arctic Canada, from Yukon Territory to Baffin Island, with the Mackenzie River delta and adjacent parts being of special importance, as well as Canada's Arctic islands north as far as Victoria and Banks Islands. Its associated migration routes (see Map 4) extend north-south as far as 1,700 miles from high-latitude breeding grounds as far away as Banks Island to wintering grounds that are centered from Chesapeake Bay to coastal North Carolina. More than 90 percent of the Eastern Population winters in Pennsylvania, Maryland, Virginia, and North Carolina, with the rest scattered from Ontario, Canada, to South Carolina. They are legally hunted in North Carolina and Virginia (Wilkins et al., 2010).

From 1971 to 1975 an estimated 39,000 whistling swans wintered in the Chesapeake Bay region of Maryland, Delaware, and Virginia, but by 2006–10 that average had dropped to 17,000 birds. Meanwhile, the number of swans wintering in North Carolina increased from an average of 20,700 during 1971–75 to 68,600 during 2006–10. These changes apparently were the result of pollution and high levels of nutrients damaging bay grasses along with increased sedimentation and ecological damage caused by Hurricane Agnes in 1972. Cornely, Petrie, and Hindman (2014) estimated the average 2011–13 winter survey counts of the Eastern Population to be 105,000 birds.

Historically, the Eastern Population has outnumbered the Western Population. There has been a substantial increase in whistling swan numbers in both populations over the past half-century: the Western Population estimate had increased to more than 105,000 birds by 2009, and the Eastern Population to nearly 97,000 by 2010 (Baldassarre, 2014). In 2012 the combined Eastern and Western populations numbered more than 228,000 birds (Pacific Flyway Council, 2017).

Winter whistling swan surveys in the Pacific Flyway averaged about 62,000 swans from 1955 to 2015, and 86,300 swans from 2006 to 2015. The population reached a peak of 122,521 swans in 1997. Spring waterfowl breeding surveys in Alaska (a combined index of the annual Waterfowl Breeding Population and Habitat Survey and Yukon-Kuskokwim Delta Coastal Zone Survey) averaged 109,296 swans from 1985 to 2016 and 123,426 swans from 2007 to 2016. The population reached an all-time high of 174,428 swans in 2008. Overall, the combined number of Eastern Population and Western Population swans averaged 189,798 from 2007 to 2016 and peaked at more than 228,000 tundra swans in 2012 (Pacific Flyway Council, 2017).

In 2018 the estimated Western Population totaled 152,000 birds, with no significant population trend during the previous decade, whereas the Eastern Population was estimated at 111,600 birds during the 2018 Mid-winter Survey, with an estimated annual average increase of 2 percent during the previous decade (US Fish and Wildlife Service, 2018). The long-term (1966–2017) national population trends in Audubon Christmas Bird Counts have indicated annual winter population increases of 0.33 percent in the United States and 0.36 percent in Canada (Meehan et al., 2018) (see https://www.audubon.org/conservation/where-have-all-birds-gone).

Bewick's swan status

The population of Bewick's swans that breeds in eastern Russia and winters in northwestern Europe was estimated in the early 1970s at about 6,000 to 7,000 birds. Of these, about half wintered in the Netherlands, about 1,500 in England, 500 to 1,000 in Ireland, 700 in Denmark, and 300 in West Germany. By 1990 the estimated northwestern wintering population had increased to about 29,000 birds, while the eastern Asian population totaled some 86,000 swans (Rose and Scott, 1994).

The northwestern European wintering population has declined in more recent years, for still uncertain reasons, from an estimated winter population of 29,000 in 1995 to 18,000 in 2010. Estimated annual survival rates dropped from 85.3 percent to 77.3 percent or 68.1 percent, depending on the calculation methods used (Wood et al., 2018). The restricted winter quarters of both Bewick's and whistling swans make them susceptible to future population reductions, along with the effects of such environmental threats as habitat loss, pollution, and diseases. The extensive tundra breeding grounds of the species are perhaps not yet in immediate danger, although global warming effects such as rising ocean levels and increased seasonal flooding of lowland tundras in the Arctic will be an increasing problem for all arctic birds.

The most recent estimates of the Asian nonbreeding season population of the Bewick's swan are from Jia et al. (2016). Based on combined counts from Korea, Japan, and China, the estimated total population was 99,000 to 141,000 swans. Rees et al. (2019) provided a recent collective world population estimate of 120,000 birds.

Hunting and lead poisoning influences

Regular, albeit limited and localized, "recreational" hunting seasons for whistling swans have existed in the United States since 1962, but the present hunting mortality levels of a few thousand birds have seemingly not had any significant deleterious effects on the population. Apart from Utah, where hunting has

Bewick's swans, pair with downy cygnets

been regularly allowed since 1962 (Huener, 1992), there have also been limited
hunting seasons on Western Population swans in Montana since 1970 (Herbert,
1992) and Nevada in 1969 (Retterer, 1992) as well as in North Dakota (Johnson and Kohn, 1991), South Dakota (Vaa, 1992), and Alaska. A small number
of trumpeter swans have been accidentally killed during these seasons, especially in Utah and Nevada.

Eastern Population whistling swans have been legally hunted along the Atlantic coast since 1962. Serie and Bartonek (1991a) and Serie, Luszcz, and Raftovich (2002) reviewed the demographic aspects of hunting on the Eastern Population swans in North Carolina and Virginia. Estimated kills by hunters of the
Eastern Population averaged 3,382 birds annually from 2000 to 2008, and an

average of 1,057 birds were killed annually in the Western Population from 1994 to 2010 (Baldassarre, 2014).

Approximately 4,000 whistling swans were thus killed legally each year in the United States during the early 2000s, while poachers and Native subsistence hunters killed an estimated additional 6,000 to 10,000 birds. Alaskan subsistence hunters killed a similar total of about 3,600 to 7,500 whistling swans annually between 2004 and 2013 (Pacific Flyway Council, 2017). Additionally, from 2004 to 2011, the average annual harvest of swan eggs by Alaskan Natives was 1,381 eggs, ranging from 682 to 2,607 (Naves, 2015).

Sherwood (1960) reported that 29 of 58 (50 percent) whistling swans examined in the Great Salt Lake region of Utah were found to be carrying lead shot in their body tissues, even though hunting of the species was then illegal everywhere except for Native American subsistence hunting in Alaska and Canada. Significant local mortality to trumpeter and whistling swans also occurs from the ingestion of lead shot, as for example at Whatcom County, Washington, and Sumas Prairie, British Columbia (Smith et al., 2007). Among 236 gizzards of trumpeter and whistling swans from western Washington, lead shot was found in 33 percent, and 35 percent of 110 livers had lead concentrations diagnostic of lead poisoning (Lagerquist, Davison, and Foreyt, 1994).

Although flying accidents, especially collisions with overhead lines, were found to be the most frequent cause of death among 182 dead Bewick's swans found dead around Wildfowl and Wetlands Trust centers in England, lead poisoning was determined to be the second most common cause (Rees, 2006). Although Bewick's swans are protected virtually everywhere in Europe, 34 percent of 272 swans wintering in Britain during the early 1970s were found to be carrying lead pellets in their bodies, including 44 percent of all adults. Similarly, during studies made between 1990 and 2000, 39 percent of 191 adult Bewick's swans had one or more lead pellets present, although more recent studies have indicated lower percentages of affected birds. Smaller percentages of birds with lead pellets have also been reported for whooper swans in Britain (Brazil, 2003; Rees et al., 2019).

III. References

A bibliographic introduction and history of the exhaustive technical swan literature might be helpful. In 1946 Peter Scott established Britain's now famous waterfowl research and conservation organization and center for Bewick's swan research, the Wildfowl Trust (TWT). It is situated near the Severn River and the village of Slimbridge, Gloucestershire, and was originally known as the Severn Wildfowl Trust. Its major annual technical publication was titled *The Annual Report of the Wildfowl Trust*. This publication's title was changed to *Wildfowl* in 1969, and still later the trust's title (and mission) was expanded into the Wildfowl and Wetlands Trust as part of a larger conservation effort to preserve and research wetlands. The Eurasian journal *Casarca* is a periodic research journal of an international consortium of waterfowl biologists known as the Goose, Swan, and Duck Study Group of Eastern Europe and North Asia. Five international swan conferences have been organized in various locations, followed by associated publications that contain the conference's research reports or their abstracts.

The Trumpeter Swan Society (TTSS) is a North American nonprofit organization formed in 1968 and primarily devoted to the conservation and restoration of trumpeter swans with a secondary interest in the other North American swans. From 1971 through 1996, the society published a newsletter for members. In 1991 another publication, *Trumpetings*, was produced that became the member newsletter in 1997. In 1997 TTSS initiated *North American Swans: Bulletin of the Trumpeter Swan Society*. Since 1997 summaries of TTSS's annual conference proceedings have been published as issues of *North American Swans*. A set of the proceedings is housed in the society's office at Three Rivers Park District, 12615 Rockford Road, Plymouth, Minnesota, 55441-1248, and can be ordered from the society at ttss@trumpeterswansociety/org.

The Swan Specialist Group (associated with Wetlands International and the International Union for Conservation of Nature [IUCN]) maintains a website that lists recent technical literature on swans, including electronic links to texts of original publications (see http://www.swansg.org/resources/literature/). Six International Swan Symposia have been held under the auspices of the Wetlands International/IUCN–Species Survival Commission (SSC), Swan Specialist Group: at Slimbridge, UK (1971); Sapporo, Japan (1980); Oxford, UK (1989); Warrenton, Virginia, USA (2001); Easton, Maryland, USA (2014); and Tartu, Estonia (2018). Published proceedings of most symposia have appeared, beginning with the second symposium, most recently through the journal *Waterbirds*.

Waterfowl Monographs

Baldassarre, G. 2014. *Ducks, Geese, and Swans of North America*. 2 vols. Rev. ed. Baltimore, MD: Johns Hopkins University Press. 1,027 pp.

Bauer, K. M., and U. N. Glutz von Blotztheim. 1968–69. *Handbuch der Vogel Mitteleuropas*. Bands 2 and 3. Frankfurt am Main: Akademische Verlagsgesellschaft.

Bellrose, F. C. 1976. *Ducks, Geese, and Swans of North America*. 2nd ed. Harrisburg, PA: Stackpole.

Bent, A. C. 1923. *Life Histories of North American Wild Fowl*. Part 1. US National Museum Bulletin 126. Washington, DC: US Government Printing Office.

———. 1925. *Life Histories of North American Wild Fowl*. Part 2. US National Museum Bulletin 130. Washington, DC: US Government Printing Office.

Birkhead, M., and C. Perrins. 1986. *The Mute Swan*. London, UK: Croom Helm. 157 pp.

Delacour, J. 1954–64. *The Waterfowl of the World*. 4 vols. London, UK: Country Life. (Swans are described in volume 1, 1954.)

Johnsgard, P. A. 1965. *Handbook of Waterfowl Behavior*. Ithaca, NY: Cornell University Press. 378 pp.

———. 1968. *Waterfowl: Their Biology and Natural History*. Lincoln, NE: University of Nebraska Press. 138 pp.

———. 1975. *Waterfowl of North America*. Bloomington: Indiana University Press, 575 pp. (Rev. ed., 2010: http://digitalcommons.unl.edu/biosciwaterfowlna/1)

———. 1978. *Ducks, Geese, and Swans of the World*. Lincoln: University of Nebraska Press. 404 pp. http://digitalcommons.unl.edu/biosciducksgeeseswans/

———. 1987. *Waterfowl of North America: The Complete Ducks, Geese and Swans*. (By Johnsgard (author of species descriptions), Robin Hill (artist), S. D. Ripley, and the Duke of Edinburgh. Augusta, GA: Morris. 135 pp.

———. 2016a. *Swans: Their Biology and Natural History*. Lincoln: University of Nebraska–Lincoln DigitalCommons and Zea Books. 114 pp. http://digitalcommons.unl.edu/zeabook/38/

———. 2016b. *The North American Geese: Their Biology and Behavior*. Lincoln: University of Nebraska–Lincoln DigitalCommons and Zea Books. 159 pp. http://digitalcommons.unl.edu/zeabook/44/

———. 2016c. *The North American Sea Ducks: Their Biology and Behavior*. Lincoln: University of Nebraska–Lincoln DigitalCommons and Zea Books. 256 pp. http://digitalcommons.unl.edu/zeabook/50/

———. 2017a. *The North American Perching and Dabbling Ducks: Their Biology and Behavior*. Lincoln: University of Nebraska–Lincoln DigitalCommons and Zea Books. 228 pp. http://digitalcommons.unl.edu/zeabook/53/

———. 2017b. *The North American Whistling-Ducks, Pochards and Stifftails*. Lincoln: University of Nebraska–Lincoln DigitalCommons and Zea Books. 188 pp. http://digitalcommons.unl.edu/zeabook/54/

Kear, J. 2005. *Ducks, Geese, and Swans*. 2 vols. Oxford, UK: Oxford University Press. 910 pp.

Kortright, F. H. 1942. *The Ducks, Geese, and Swans of North America: A Vade Mecum for the Naturalist and the Sportsman*. Washington, DC: American Wildlife Institute.

Madge, S. C., and H. Burn. 1988. *Waterfowl: An Identification Guide to the Ducks, Geese, and Swans of the World*. Boston: Houghton Mifflin.

Matthews, G. V. T., and M. Smart, eds. 1981. *Proceedings of the Second International Swan Symposium, Sapporo, Japan, February 21–22, 1980*. Slimbridge, UK: International Waterfowl Research Bureau (IWRB).

McCoy, J. J. 1967. *Swans*. New York: Lothrop, Lee & Shepard. 160 pp.

Owen, M. 1977. *Wildfowl of Europe*. London: Macmillan.

Owen, M., G. L. Atkinson-Willes, and D. G. Salmon. 1986. *Wildfowl in Great Britain*. 2nd ed. Cambridge, UK: Cambridge University Press.

Palmer, R. S., ed. 1976. *Handbook of North American Birds*. Vols. 2 and 3. New Haven, CT: Yale University Press.

Scott, P., and Boyd, H. 1957. *Wildfowl of the British Isles*. London: Country Life.

Sears, J., and P. J. Bacon, eds. 1991. *Wildfowl: Supplement Number 1, Third IWRB International Swan Symposium, Oxford, England, December 9–13, 1989*. Slimbridge, UK: International Waterfowl and Wetlands Research Bureau.

Todd, F. S. 1979. *Waterfowl: Ducks, Geese, and Swans of the World*. New York: Harcourt Brace Jovanovich.

———. 1996. *Natural History of the Waterfowl*. Vista, CA: Ibis Press. 490 pp.

———. 2018. *North American Ducks, Geese, and Swans: Identification Guide*. Surrey, BC: Hancock House.

Other Waterfowl and General Bird References

American Ornithologists' Union. 1998. *Check-list of North American Birds*. 7th ed. Washington, DC: American Ornithologists' Union. (Plus annual supplements through 2019.)

Armstrong, R. H. 2015. *Guide to the Birds of Alaska*. 6th ed. Portland, OR: Graphic Art Books. 268 pp.

Atkinson-Willes, G., ed. 1963. *Wildfowl in Great Britain: A Survey of the Winter Distribution of the Anatidae and Their Conservation in England, Scotland, and Wales*. Monographs of the Nature Conservancy, no. 3. London: HMSO.

Bannerman, D. A. 1957. *The Birds of the British Isles*. Vol. 6. London: Oliver & Boyd.

BirdLife International. 2004. *Birds in Europe: Population Estimates, Trends, and Conservation Status*. Cambridge, UK: BirdLife International (BirdLife Conservation Series No. 12).

———. 2015. *European Red List of Birds*. Available at https://www.birdlife.org/europe-and-central-asia/european-red-list-birds-0

Bottler, P. D. 1983. *Systematic Relationships among the Anatidae: An Immunological Study with a History of Anatid Classification and a System of Classification*. PhD diss. New Haven, CT: Yale University.

Brush, A. H. 1976. Waterfowl feather proteins: Analysis of use in taxonomic studies. *Journal of Zoology* 179: 467–498.

Campbell, R. W., N. K. Dawe, I. McTaggart-Cowan, J. M. Cooper, G. W. Kaiser, and M. C. E. McNall. 1990. *The Birds of British Columbia, Volume 1: Nonpasserines, Introduction and Loons through Waterfowl*. Vancouver: University of British Columbia Press.

Clements, J. F. 2007. *The Clements Checklist of the Birds of the World*. 6th ed. Ithaca, NY: Cornell University Press.

Cramp, S., K. E. L. Simmons, et al., eds. 1977. *Handbook of the Birds of Europe, the Middle East, and North Africa: The Birds of the Western Palearctic, Vol. 1: Ostrich to Ducks*. Oxford, UK: Oxford University Press. 722 pp.

Delacour, J., and E. Mayr. 1945. The family Anatidae. *Wilson Bulletin* 57: 3–55.

del Hoyo J., A. Elliott, and J. Sargatal. 1992. *Handbook of the Birds of the World*. Vol. 1. Barcelona: Lynx Editions. 696 pp.

Dementiev, G. P., & N. A. Gladkov, eds. 1967. *Birds of the Soviet Union*. Vol. 2. Washington, DC: Israel Program for Scientific Translations, US Department of the Interior and National Science Foundation.

Dickinson, E., ed. 2013. *The Howard and Moore Complete Checklist of the Birds of the World, Volume 1: Non-Passerines*. Eastbourne, UK: Aves Press.

Gabrielson, I. N., and F. C. Lincoln. 1959. *The Birds of Alaska*. Washington, DC: Wildlife Management Institute, and Harrisburg, PA: Stackpole.

Godfrey, W. E. 1986. *The Birds of Canada*. Rev. ed. Ottawa: National Museum of Natural Sciences.

González, J., H. Düttmann, and M. Wink. 2009. Phylogenetic relationships based on two mitochondrial genes and hybridization patterns in Anatidae. *Journal of Zoology* 279: 310–318.

Harshman, J. 1996. Phylogeny, evolutionary rates, and ducks. PhD diss. Chicago: University of Chicago.

Hickey, J. J. 1952. *Survival Studies of Banded Birds*. Special Scientific Report: Wildlife, No. 15. Washington, DC: US Department of the Interior, Fish and Wildlife Service. 177 pp.

Hudson, R., ed. 1975. *Threatened Birds of Europe*. London: Macmillan.

Jacob, J., and A. Glaser. 1975. Chemotaxonomy of Anseriformes. *Biochemical Systematics and Ecology* 2: 215–220.

Jia, Q., K. Koyama, C.-Y. Choi, H.-J. Kim, L. Cao, D. Gao, G. Liu, and A. D. Fox. 2016. Population estimates and geographical distributions of swans and geese in East Asia based on counts during the non-breeding season. *Bird Conservation International* 26(4): 397–417.

Johnsgard, P. A. 1960. Hybridization in the Anatidae and its taxonomic implications. *Condor* 62: 25–33. http://digitalcommons.unl.edu/biosciornithology/71

———. 1961a. The taxonomy of the Anatidae—a behavioural analysis. *Ibis* 103: 71–85. http://digitalcommons.unl.edu/johnsgard/29

———. 1961b. Tracheal anatomy of the Anatidae and its taxonomic significance. *Wildfowl* 12: 58–69.

———. 1962. Evolutionary trends in the behaviour and morphology of the Anatidae. *Wildfowl* 13: 130–148.

———. 1963. Behavioral isolating mechanisms in the family Anatidae. Pp. 531–543 in *Proceedings XIII International Ornithological Congress, Ithaca, New York, June 17–24, 1962* (Charles G. Sibley, ed.). Baton Rouge, LA: American Ornithologists' Union. http://digitalcommons.unl.edu/johnsgard/23

———. 1972. Observations on sound production of the Anatidae. *Wildfowl* 22: 46–59.

———. 1979. Order Anseriformes (Anatidae and Anhimidae). Pp. 425–506 in *Check-List of the Birds of the World*, 2nd ed. (E. Mayr and G. W. Cottrell, eds.). Cambridge, MA: Harvard University Press. http://digitalcommons.unl.edu/johnsgard/32

———. 2012a. *Wetland Birds of the Central Plains: South Dakota, Nebraska, and Kansas*. Lincoln: University of Nebraska–Lincoln DigitalCommons and Zea Books. 276 pp. http://digitalcommons.unl.edu/zeabook/8

———. 2012b. *Wings Over the Great Plains: Bird Migrations in the Central Flyway*. Lincoln: University of Nebraska-Lincoln DigitalCommons and Zea Books. 245 pp. http://digitalcommons.unl.edu/zeabook/13

Kessel, B., and D. G. Gibson. 1978. *Status and Distribution of Alaska Birds*. Studies in Avian Biology No. 1. Los Angeles: Cooper Ornithological Society.

Kolbe, H. 1972. *Die Entenvogel der Welt*. Redebuel, Germany: Neumann Verlag.

Lack, D. 1967. The significance of clutch size in waterfowl. *Wildfowl Trust Annual Report* 18: 125–128.

———. 1968a. The proportion of yolk in the eggs of waterfowl. *Wildfowl* 19: 67–69.

———. 1968b. *Ecological Adaptations for Breeding in Birds*. London: Methuen & Co.

———. 1974. *Evolution Illustrated by Waterfowl*. Oxford, UK: Blackwell Scientific Publications.

Livezey, B. C. 1996. A phylogenetic analysis of geese and swans (Anseriformes: Anserinae), including selected fossil species. *Systematic Biology* 45: 415–450.

Lutmerding, J. A., and A. S. Love. 2011. *Longevity Records of North American Birds*. Version 2011.2. Laurel, MD: Bird Banding Laboratory, Patuxent Wildlife Research Center.

Madsen, C. S., K. P. McHough, and S. R. De Kloet. 1988. A partial classification of waterfowl (Anatidae) based on single-copy DNA. *Auk* 105: 452–459.

Martin, A., H. S. Zim, and A.L. Nelson. 2011. *American Wildlife and Plants: A Guide to Wildlife Food Habits*. Reprint of 1951 edition. New York: Dover.

McKinney, F. 1965b. The comfort movements of Anatidae. *Behaviour* 25: 120–220.

Meehan, T. D., G. S. LeBaron, K. Dale, N. L. Michel, G. M. Verutes, and G. M. Langham. 2018. *Abundance Trends of Birds Wintering in the USA and Canada, from Audubon Christmas Bird Counts, 1966–2017*. Version 2.1. New York: National Audubon Society,

Murie, O. J. 1959. *Fauna of the Aleutian Islands and Alaska Peninsula*. Washington, DC: US Fish and Wildlife Service, North American Fauna, No. 61: 1–406.

Murton, R. K., and J. Kear. 1978. Photoperiodism in waterfowl: Phasing of breeding cycles and zoogeography. *Journal of Zoology* 186: 243-283.

Murton, R. K., and N. J. Westwood. 1977. *Avian Breeding Cycles*. Oxford, UK: Oxford University Press.

Nelson, A. D., and Martin, A. C. 1953. Gamebird weights. *Journal of Wildlife Management* 17: 36-42.

Olson, S. M. 2018. *Pacific Flyway Data Book 2018*. Vancouver, WA: US Department of the Interior, Fish and Wildlife Service, Division of Migratory Bird Management. https://www.fws.gov/migratory-birds/pdf/surveys-and-data/DataBooks/PacificFlywayDatabook.pdf

Pardieck, K. L., D. J. Ziolkowski Jr., M. Lutmerding, V. Aponte, and M-A. R. Hudson. 2019. North American Breeding Bird Survey Dataset 1966-2018. Version 2018.0. Laurel, MD: Patuxent Wildlife Research Center. https://doi.org/10.5066/P9HE8XYJ

Parkes, K. C. 1958. Systematic notes on North American birds: 2. The waterfowl (Anatidae). *Annals of the Carnegie Museum* 35: 117-125.

Ploeger, P. L. 1968. Geographical differentiation in arctic Anatidae as a result of isolation during the last glacial. *Ardea* 56: 1-159.

Raftovich, R. V., K. K. Fleming, S. C. Chandler, and C. M. Cain. 2019. *Migratory Bird Hunting Activity and Harvest during the 2017-18 and 2018-19 Hunting Seasons*. Laurel, MD: US Fish and Wildlife Service.

Richards, J. M., and A. J. Garston, eds. 2018. *Birds of Nunavut*. 2 vols. Vancouver: University of British Columbia Press. 820 pp.

Rohwer, F. C., and D. I. Eisenhauer. 1989. Egg mass and clutch size relationships in geese, eiders, and swans. *Ornis Scandinavica* 20: 43-48.

Sears, J., and P. J. Bacon, eds. 1991. *Wildfowl: Supplement Number 1, Third IWRB International Swan Symposium, Oxford, England, December 9-13, 1989*. Slimbridge, UK: International Waterfowl and Wetlands Research Bureau.

Sibley, C. G., and B. L. Monroe Jr. 1990. *Distribution and Taxonomy of Birds of the World*. New Haven, CT: Yale University Press.

Sibley, C. G., and J. E. Ahlquist. 1990. *Phylogeny and Classification of Birds: A Study in Molecular Evolution*. New Haven, CT: Yale University Press.

Snyder, L. L. 1957. *Arctic Birds of Canada*. Toronto: University of Toronto Press.

Sun, Z., T. Pan, C. Hu, L. Sun, H. Ding, H. Wang, C. Zhang, H. Jin, Q. Chang, X. Kan, and B. Zhang. 2017. Rapid and recent diversification patterns in Anseriformes birds: Inferred from molecular phylogeny and diversification analyses. *PloS One* 12(9): e0184529.

Trumpeter Swan Society and Wetlands International. 2014. *Program Schedule and Presentation Abstracts, 23rd Swan Conference and 5th International Swan Conference, February 3-6, 2014, Easton, Maryland*. 61 pp. https://www.trumpeterswansociety.org/swan-information/swan-library/swan-library/ttss-swan-conference-abstracts.html

US Fish and Wildlife Service (USFWS). 1954. *Waterfowl Populations and Breeding Conditions, Summer 1953*. Washington, DC: US Fish and Wildlife Service, Special Scientific Report: Wildlife, No. 25. 294 pp.

———. 2015. *Productivity Surveys of Geese, Swans, and Brant Wintering in North America, 2009*. Arlington, VA: Division of Migratory Bird Management, US Fish and Wildlife Service. 55 pp.

———. 2018. *Waterfowl Population Status, 2018*. Washington, DC: Division of Migratory Bird Management, US Fish and Wildlife Service.

Voous, K. H. 1960. *Atlas of European Birds*. London: Elsevier, Nelson.

Weller, M. W. 1964a. General habits. Pp. 15-34 in *The Waterfowl of the World*, vol. 4 (J. Delacour, ed.). London: Country Life.

———. 1964b. The reproductive cycle. Pp. 35-79 in *The Waterfowl of the World*, vol. 4 (J. Delacour, ed.). London: Country Life.

———. 1964c. Ecology. Pp. 80-107 in *The Waterfowl of the World*, vol. 4 (J. Delacour, ed.). London: Country Life.

———. 1964d. Distribution and species relationships. Pp. 108–120 in *The Waterfowl of the World*, vol. 4 (J. Delacour, ed.). London: Country Life.

Wetlands International (compiled by S. Delany and D. Scott). 2006. *Waterbird Population Estimates*. 4th ed. Wageningen, The Netherlands: Wetlands International. 239 pp.

Woolfenden, G. E. 1961. Postcranial osteology of the waterfowl. *Bulletin of the Florida State Museum, Biological Sciences* 6: 1–29.

Major Swan References

Banko, W. 1960. *The Trumpeter Swan: Its History, Habits, and Population in the United States*. Washington, DC: US Department of the Interior, Fish and Wildlife Service, North American Fauna No. 63. 214 pp.

Banko, W. E., and A. W. Schorger. 1976. Trumpeter swan. Pp. 5–71 in *Handbook of North American Birds*, vol. 2 (Waterfowl, pt. 1) (R. S. Palmer, ed.). New Haven, CT: Yale University Press.

Banko, W. E., and R. H. Mackay. 1964. Our native swans. Pp. 155–164 in *Waterfowl Tomorrow* (J. P. Linduska, ed.). Washington, DC: US Department of the Interior, Bureau of Sport Fisheries and Wildlife.

Bart, J., S. L. Earnst, and P. J. Bacon. 1991. Comparative demography of the swans: A review. Pp. 15–21 in *Wildfowl: Supplement Number 1, Third IWRB International Swan Symposium, Oxford, England, December 9–13, 1989* (J. Sears and P. J. Bacon, eds.). Slimbridge, UK: International Waterfowl and Wetlands Research Bureau.

Blus, L. J. 1994. A review of lead poisoning in swans. *Comparative Biochemistry and Physiology* 108C(3): 259–267.

Brazil, M. 2003. *The Whooper Swan*. London: T & AD Poyser. 512 pp.

Boyd, H. 1972. Classification. Pp. 17–28 in *The Swans* (P. Scott, ed.). Boston: Houghton Mifflin.

Ciaranca, M. A., C. C. Allin, and G. S. Jones. 1997. Mute swan (*Cygnus olor*). Version 2.0. *Birds of North America* (A. F. Poole and F. B. Gill, eds.). Ithaca, NY: Cornell Laboratory of Ornithology. https://birdsna.org/Species-Account/bna/species/mutswa/introduction

Hilprecht. A. 1956. *Höcherschwan, Singshwan, Zwergschwan*. Wittenberg: Neue Brehm Bücherei.

Johnsgard, P. A. 2016. *Swans: Their Biology and Natural History*. Lincoln: University of Nebraska-Lincoln DigitalCommons and Zea Books. 114 pp. http://digitalcommons.unl.edu/zeabook/38/

Limpert, R. J., and S. L. Earnst. 1994. Tundra swan (*Cygnus columbianus*). Version 2.0. *Birds of North America* (A. F. Poole and F. B. Gill, eds.). Ithaca, NY: Cornell Laboratory of Ornithology. https://birdsna.org/Species-Account/bna/species/tunswa/introduction

Livezey, B. C. 1986. A phylogenetic analysis of recent anseriform genera using morphological characters. *Auk* 103: 737–754.

———. 1996. A phylogenetic analysis of the geese and swans (Anseriformes, Anserinae). *Systematic Biology* 45(4): 415–450.

Ma, M., and Cai, D. 2000. *Swans in China*. Maple Plain, MN: The Trumpeter Swan Society. (First published in Chinese: M. Ma and D. Cai, *Wild Swans* [Beijing: China Meteorological Press, 1993].)

Matthews, G. V. T., and M. Smart, eds. 1981. *Proceedings of the Second International Swan Symposium, Sapporo, Japan, February 21–22, 1980*. Slimbridge, UK: International Waterfowl Research Bureau (IWRB).

Mitchell, C. D., and M. W. Eichholz. 2019. Trumpeter swan (*Cygnus buccinator*). Version 3.0. *Birds of North America* (A. F. Poole, ed.). Ithaca, NY: Cornell Laboratory of Ornithology. https://birdsna.org/Species-Account/bna/species/truswa/introduction

Ogilvie, M. A. 1972. Distribution, numbers, and migration. Pp. 29–56 in *The Swans* (P. Scott, ed.). Boston: Houghton Mifflin.

Paca, L. G. 1963. *The Royal Birds*. New York: St. Martin's Press. 164 pp.

Petzold, H.-G. 1964. Beiträge zur vergleichenden Ethologie der Schwane (Anseres, Anserini). *Beiträge zur Vogelkunde* 10: 1–123.

Price, A. L. 1995. *Swans of the World: In Nature, History, Myth, and Art*. Tulsa, OK: Council Oak Distribution. 176 pp.

Rees, E. C. 2006. *Bewick's Swan*. London: T. & A. D. Poyser. 296 pp.

Rees, E. C., S. L. Earnst, and J. Coulson, eds. 2002. Special Publication 1: Proceedings of the Fourth International Swan Symposium 2001. *Waterbirds* 25.

Schull, M. 2012. *The Swan: A Natural History*. Ludlow, UK: Merlin Unwin Books. 224 pp.

Scott, D. 1995. *Swans*. Minneapolis, MN: Voyageur Press. 72 pp.

Scott, P., ed., and the Wildfowl Trust. 1972. *The Swans*. London: Michael Joseph.

Sears, J., and P. J. Bacon, eds. 1991. *Wildfowl Supplement Number 1, Third IWRB International Swan Symposium, Oxford, England, December 9–13, 1989*. Slimbridge, UK: International Waterfowl and Wetlands Research Bureau.

Stejneger, L. 1882. Outline of a monograph of the Cygninae. *Proceedings of the US National Museum* 5: 174–221.

Ticehurst, N. F. 1967. *The Mute Swan in England*. London: Cleaver-Hume Press.

Wilmore, S. B. 1974. *Swans of the World*. New York: Taplinger.

Young, P. 2008. *Swan*. London: Reaktion Books.

Multispecies Swan Studies

Albertsen, J. O., and Y. Kamazawa. 2002. Numbers and ecology of swans wintering in Japan. Special Publication 1: Proceedings of the Fourth International Swan Symposium 2001 (E. C. Rees, S. L. Earnst and J. Coulson, eds.). *Waterbirds* 25: 74–85.

Atkinson-Willes, G. 1981. The numerical distribution and the conservation requirements of *Cygnus olor, Cygnus cygnus*, and *Cygnus columbianus bewickii* in north-west Europe. Pp. 40–48 in *Proceedings of the Second International Swan Symposium, Sapporo, 1980* (G. V. T. Matthews and M. Smart eds.). Slimbridge, UK: International Waterfowl Research Bureau.

Bart, J., S. Earnst, and P. J. Bacon. 1991. Comparative demography of the swans: A review. *Wildfowl* (Supplement 1): 15–21.

Bartonek, J. C., W. W. Blandin, K. E. Gamble, and H. W. Miller. 1981. Numbers of swans wintering in the United States. Pages 19–25 in *Proceedings of Second International Swan Symposium, Sapporo, 1980* (G. V. T. Matthews and M. Smart, eds.). Slimbridge, UK: International Waterfowl and Wetlands Research Bureau.

Black, J. M. 1988. Preflight signaling in swans: A mechanism for group cohesion and flock formation. *Ethology* 79: 143–157.

Bollinger, K. S. 1982. *Progress Report on Behavior of Nesting Trumpeter and Tundra Swans at Minto Flats, Alaska—1982*. Fairbanks, AK: US Fish and Wildlife Service.

Boyd, S. 1994. Abundance patterns of trumpeter swans and tundra swans on the Fraser River, British Columbia (abstract). *Trumpeter Swan Society Conference* 14: 48.

Brown, M. J., E. Linton, and E. C. Rees. 1992. Causes of mortality among wild swans in Britain. *Wildfowl* 43: 70–79.

Bryant, J. M., B. D. Scotton, and M. R. Hans. 2005. *Sympatric Nesting Range of Trumpeter and Tundra Swans on the Koyukuk National Wildlife Refuge in Northwest Interior Alaska*. Galena, AK: US Fish and Wildlife Service, Koyukuk/Nowitna National Wildlife Refuge Complex.

Canniff, R. S. 1990. Trumpeter and tundra swan collar sightings in the Skagit Valley, 1977–1978 to 1987–1988. *Trumpeter Swan Society Conference* 11: 125–141.

Carey, C. G. 2000. Mute swan control and trumpeter swan experimental breeding project in urban central Oregon. *Trumpeter Swan Society Conference* 17: 114–116.

Central Flyway Council. 1991. Position statement on tundra swan hunting in the Central Flyway relative to potential conflicts with trumpeter swan restoration. *Trumpeter Swan Society Conference* 12: 73–74.

Conant, B., J. I. Hodges, and J. G. King. 1991. Continuity and advancement of trumpeter swan *Cygnus buccinator* and tundra swan *Cygnus columbianus* population monitoring in Alaska. *Wildfowl* (Supplement No. 1): 125–136

Cooper, B. A., and R. J. Ritchie. 1990. Migration of trumpeter and tundra swans in east-central Alaska during spring and fall, 1987. *Trumpeter Swan Society Conference* 11: 82–91.

Degernes, L., S. Heilman, M. Trogdon, M. Jordan, M. Davison, D. Kraege, M. Correa, and P. Cowen. 2006. Epidemiologic investigation of lead poisoning in trumpeter and tundra swans in Washington state, USA, 2000–2002. *Journal of Wildlife Diseases* 42: 345–358.

Drewien, R. C., J. T. Herbert, and T. W. Aldrich. 2000. Detecting trumpeter swans harvested in tundra swan hunts (abstract). *Trumpeter Swan Society Conference* 17: 155.

Froelich, A. J., J. C. Johnson, and D. M. Lodge. 1999. Food preferences of mute and trumpeter swans. *Trumpeter Swan Society Conference* 16: 133.

Harvey, N. G. 1998. Evolutionary and population genetics of swans. *Swan Specialist Group Newsletter* 7: 15–16.

———. 1999. A hierarchical genetic analysis of swan relationships. PhD diss. Nottingham, UK: University of Nottingham.

Howie, R. R., and R. G. Bison. 2004. Wintering trumpeter and tundra swans in the southern interior of British Columbia. *Trumpeter Swan Society Conference* 19: 16–28.

Janssen, R. B. 2003. Trumpeter swan–tundra swan interaction. *Loon* 75(3): 171–172.

Jobes, C. R. 1986. Energetics of growth of trumpeter and mute swan cygnets (abstract). *Trumpeter Swan Society Conference* 9: 119.

Johnsgard, P. A. 1974. The taxonomy and relationships of the northern swans. *Wildfowl* 25: 155–161. https://digitalcommons.unl.edu/johnsgard/11/

———. 2013. The swans of Nebraska. *Prairie Fire*, January 2013, pp. 12–13. http://www.prairiefire-newspaper.com/2013/01/the-swans-of-nebraska

Johnson, W. C. 1999. Observations of territorial conflict between trumpeter swans and mute swans in Michigan. *Trumpeter Swan Society Conference* 16: 134–136.

King, J. C. 1970. The swans and geese of Alaska's Arctic Slope. *Wildfowl* 21: 1–17.

Krivtsov, S. K., and Y. N. Mineev. 1991. Daily time and energy budgets of whooper swans *Cygnus cygnus* and Bewick's swans *Cygnus bewickii* in the breeding season. Pp. 319–321 in *Wildfowl: Supplement Number 1, Third IWRB International Swan Symposium, Oxford, England, December 9–13, 1989* (J. Sears and P. J. Bacon, eds.). Slimbridge, UK: International Waterfowl and Wetlands Research Bureau.

Lagerquist, J. E., and D. M. Foreyt. 1994. Lead poisoning and other causes of mortality in trumpeter (*Cygnus buccinator*) and tundra (*C. columbianus*) swans in western Washington. *Journal of Wildlife Diseases* 30(1): 60–64.

Laubek, B. 1995. Whooper swans *Cygnus cygnus* and Bewick's swans *Cygnus columbianus bewickii* wintering in Denmark: Increasing agricultural conflicts. *Wildfowl* 46: 8–15.

Loranger, A., and D. Lons. 1990. Relative abundance of sympatric trumpeter and tundra swan populations in west-central interior Alaska. *Trumpeter Swan Society Conference* 11: 92–98.

Lumsden, H. G. 1984. The pre-settlement breeding distribution of trumpeter, *Cygnus buccinator*, and tundra swans, *C. columbianus columbianus*, in eastern Canada. *Canadian Field-Naturalist* 98: 415–424.

———. 1986. The trumpeter swan/mute swan experiment: Ontario. *Trumpeter Swan Society Conference* 9: 117–118.

———. 2016. Trumpeter swans and mute swans compete for space in Ontario. *Ontario Birds* 34: 14–23.

Perfiliev, V. I. 1987. Whooper swan and Bewick's swan in Northern Yakutia. Pp. 134–135 in *Ecology and Migrations of Swans in USSR* (E. V. Syroechkovski, ed.). Moscow: Nauka.

Rees, E. C., J. M. Bowler, and J. H. Beekman. 1997. *Cygnus columbianus* Bewick's swan and whistling swan. *Birds of the Western Palearctic* (BWP Update 1): 63–74.

Rees, E. C., P. Clausen, L. Cao, and J. T. Coleman. 2019. Conservation status of the world's swan populations, *Cygnus* sp. and *Coscoroba* sp.: A review of current trends and gaps in knowledge. *Wildfowl* (Special Issue 5): 35–72.

Rees, E. C., P. Lievesley, R. A. Pettifor, and C. Perrins. 1996. Mate fidelity in swans: an interspecific comparison. Pp. 118–137 in *Partnerships in Birds: the Study of Monogamy* (J. M. Black, ed.). London: Oxford University Press.

Roy, V. 1996. Trumpeter and tundra swans: Their history and future at the Bear River Migratory Bird Refuge. *Trumpeter Swan Society Conference* 15: 53–61.

Shirkley, B., D. Luukkonen, and S. R. Winterstein. 2014. Mute and tundra swan distribution and abundance on Lake St. Clair and western Lake Erie (abstract). *Trumpeter Swan Society Conference* 23 (unpaged).

Smith, M. C., J. M. Grassley, C. E. Grue, M. Davison, J. Bohannon, C. Schneider, and L. Wilson. 2007. Mortality of swans due to ingestion of lead shot, Whatcom County, Washington, and Sumas Prairie, British Columbia. *Trumpeter Swan Society Conference* 20: 114–116.

Syroechkovski, E. E. 2002. Distribution and population estimates for swans in the Siberian arctic in the 1990s. Special Publication 1: Proceedings of the Fourth International Swan Symposium 2001 (E. C. Rees, S. L. Earnst, and J. Coulson, eds.). *Waterbirds* 25: 100–113.

US Fish and Wildlife Service. 1978. *Distribution and Abundance of Swans in Alaska* (map and text). Anchorage, AK: US Fish and Wildlife Service.

Wetlands International. 2019. *Waterbird Population Estimates*. Ede, The Netherlands: Wetlands International. http://wpe.wetlands.org/

Wetmore, A. 1951. Observations on the genera of the swans. *Journal of the Washington Academy of Sciences* 41: 338–340.

Wilk, R. J. 1993. Observations on sympatric tundra, *Cygnus columbianus*, and trumpeter swans, *C. buccinator*, in north-central Alaska, 1989–1991. *Canadian Field-Naturalist* 107: 64–68.

Wood, T., T. Brooks, and W. Sladen. 2002. Vocal characteristics of trumpeter and tundra swans and their hybrid offspring. Special Publication 1: Proceedings of the Fourth International Swan Symposium 2001 (E. C. Rees, S. L. Earnst, and J. Coulson, eds.). *Waterbirds* 25: 360–362.

Individual Swan Species References

Mute Swan

Absolom, A. F., and C. M. Perrins 1999. Double-brooded mute swans. *British Birds* 92: 365–366.

Allin, C. C. 1981. Mute swans in the Atlantic flyway. Pp. 149–154 in *Proceedings of the Second International Swan Symposium* (G. V. T. Matthews and M. Smart, eds.). Slimbridge, UK: International Waterfowl and Wetland Research Bureau.

Allin, C. C., and T. P. Husband. 2003. Mute swan (*Cygnus olor*) impact on submerged aquatic vegetation and macroinvertebrates in a Rhode Island coastal pond. *Northeastern Naturalist* 10: 305–318.

———. 2004. An evaluation of 22 years of mute swan management in Rhode Island. Pp. 19–22 in *Mute Swans and Their Chesapeake Bay Habitats: Proceedings of a Symposium* (M. C. Perry, ed.). US Geological Survey, Biological Resources Discipline Information and Technology Report USGS/BRD/ITR—2004-0005. 60 pp. . https://digitalcommons.unl.edu/usgspubs/138/

Allin, C. C., G. C. Chasko, and T. P. Husband. 1987. Mute swans in the Atlantic Flyway: A review of the history, population growth, and management needs. *Transactions of the Northeast Section of the Wildlife Society* 44: 32–47.

Arsnoe, D., and A. Duffiney. 2018. From beauty to beast. Managing mute swans in Michigan to protect native resources. *The Wildlife Professional* 12: 40–44.

Auld, J. R., C. M. Perrins, and A. Charmantier. 2013. Who wears the pants in a mute swan pair? Deciphering the effects of male and female age and identity on breeding success. *Journal of Animal Ecology* 82: 826–835.

Bacon, P. J. 1980. Status and dynamics of a mute swan population near Oxford between 1976 and 1978. *Wildfowl* 31: 37–50.

———. 1980a. A possible advantage for the Polish morph of the mute swan. *Wildfowl* 31: 51–52.

———. 1980b. Status and dynamics of a mute swan population near Oxford between 1976 and 1978. *Wildfowl* 31: 37–50.

Bacon, P. J., and A. E. Coleman. 1986. An analysis of weight changes in the mute swan *Cygnus olor*. *Bird Study* 33: 145–158.

Bacon, P. J., and P. Andersen-Harild. 1989. Mute swan. Pp. 347–386 in *Lifetime Reproduction in Birds* (I. Newton, ed.). San Diego, CA: Academic Press.

Badzinski, S. S. 2008. Mute swan (*Cygnus olor*). Pp. 64–65 in *Atlas of the Breeding Birds of Ontario, 2001-2005* (M. D. Cadman, D. A. Sutherland, G. G. Beck, D. Lepage, and A. R. Couturier, eds.). Published by Bird Studies Canada, Environment Canada/Canadian Wildlife Service, Ontario Field Ornithologists, Ontario Ministry of Natural Resources, and Ontario Nature.

Badzinski, S. S., C. D. Ankney, and S. A. Petrie. 2006. Influence of migrant tundra swans (*Cygnus columbianus*) and Canada geese (*Branta canadensis*) on aquatic vegetation at Long Point, Lake Erie, Ontario. *Hydrobiologia* 567: 195–211.

Bailey, M., S. A. Petrie, and S. S. Badzinski. 2008. Diet of mute swans in Lower Great Lakes coastal marshes. *Journal of Wildlife Management* 72: 726–732.

Berglund, B. E., K. Curry-Lindahl, H. Luther, V. Olsson, W. Rodke, and G. Sellerberg. 1965. Ecological studies on the mute swan (*Cygnus olor*) in southeastern Sweden. *Acta Vertebratica* 2(2): 1–120.

Birkhead, M. 1982. Cause of mortality in the mute swan *Cygnus olor* on the River Thames. *Journal of Zoology* (London) 198: 15–25.

Birkhead, M., and C. Perrins. 1986. *The Mute Swan*. London: Croom Helm.

Birkhead, M. E., P. J. Bacon, and P. Walter. 1983. Factors affecting the breeding success of the mute swan. *Journal of Animal Ecology* 52: 727–741.

Black, J. M. 1986. Mute swans foot-slapping as a territorial advertisement display. *British Birds* 79: 500–501.

Boase, H. 1959. Notes on the display, nesting, and moult of the mute swan. *British Birds* 52: 114–121.

Breault, A. 2004. Status and management of mute swans in southwest British Columbia (abstract). *Trumpeter Swan Society Conference* 19: 203.

Chesapeake Bay Mute Swan Working Group. 2004. *Mute Swan* (Cygnus olor) *in the Chesapeake Bay: A Bay-Wide Management Plan*. Annapolis: Maryland Department of Natural Resources. https://dnr.maryland.gov/wildlife/Documents/Mute_Swan_Chesapeake_Bay_Plan_2005.pdf

Ciaranca, M. 1990. Interactions between mute swans (*Cygnus olor*) and native waterfowl in southeastern Massachusetts on freshwater ponds. MS thesis. Boston: Northeastern University.

Ciaranca, M. A., C. C. Allin, and G. S. Jones. 1997. Mute swan (*Cygnus olor*). Version 2.0. *Birds of North America* (A. F. Poole and F. B. Gill, eds.). Ithaca, NY: Cornell Laboratory of Ornithology. https://birdsna.org/Species-Account/bna/species/mutswa/introduction

Cobb, J. S., and M. M. Harlan. 1980. Mute swan (*Cygnus olor*) feeding and territoriality affects diversity and density of rooted aquatic vegetation. *American Zoologist* 20: 882.

Coleman, A. E., and C. D. T. Minton. 1979. Pairing and breeding of mute swans in relation to natal area. *Wildfowl* 30: 27–30.

———. 1980. Mortality of mute swan progeny in an area of south Staffordshire. *Wildfowl* 31: 22–28.

Coleman, A. E., J. T. Coleman, P. A. Coleman, and C. D. T. Minton. 2001. A 39-year study of a mute swan *Cygnus olor* population in the English Midlands. *Ardea* 89 (Special Issue):123–133.

Coleman, J. T., A. E. Coleman, and D. Elphick. 1994. Incestuous breeding in the mute swan *Cygnus olor*. *Ringing & Migration* 15: 127–128.

Conover, M. R., and G. S. Kania. 1994. Impact of interspecific aggression and herbivory by mute swans on native waterfowl and aquatic vegetation in New England. *Auk* 111: 744–748.

Costanzo, C., C. Davies, M. DiBona, J. Fuller, L. Hindman, M. Huang, J. Lefebvre, T. Nichols, J. Osen-kowski, P. Padding, C. Poussart, E. Reed, and D. Sausville. 2015. *Atlantic Flyway Mute Swan Management Plan*. Snow Goose, Brant, and Swan Committee, US Fish and Wildlife Service. 30 pp.

Czapulak, A. 2002. Timing of the primary moult in breeding mute swans. Special Publication 1: Proceedings of the Fourth International Swan Symposium 2001 (E. C. Rees, S. L. Earnst, and J. Coulson, eds.). *Waterbirds* 25: 258-267.

de Haan, Y. 2009. Reproductive energetics of mute swans, *Cygnus olor*, on the lower Great Lakes. Honors thesis. London: University of Western Ontario.

Demarest, J. 1981. Seasonal variation, sex difference, and habituation of territorial behavior in *Cygnus olor*. Pp. 312-318 in *Proceedings of the Second International Swan Symposium* (G. V. T. Matthews and M. Smart, eds.). Slimbridge, UK: International Waterfowl and Wetland Research Bureau.

Ellis, M. M., and C. S. Elphick. 2007. Using a stochastic model to examine the ecological, economic, and ethical consequences of population control in a charismatic invasive species: Mute swans in North America. *Journal of Applied Ecology* 44: 312-322.

Fenwick, G. H. 1983. Feeding behavior of waterfowl in relation to changing food resources in the Chesapeake Bay. PhD diss. Baltimore, MD: Johns Hopkins University.

Gayet, G., M. Guillemain, M. Benmergui, F. Mesleard, T. Boulinier, J-P. Bienvenu, H. Fritz, and J. I. Broyer. 2010. Effects of seasonality, isolation, and patch quality for habitat selection processes by mute swans *Cygnus olor* in a fish pond landscape. *Oikos* 120(6): 801-812.

Gayet, G., M. Guillemain, P. Defos du Rau, and P. Grillas. 2014. Effects of mute swan on wetlands: A synthesis. *Hydrobiologia* 723: 195-204.

Gelston, W. L., and R. D. Wood. 1982. *The Mute Swan in Northern Michigan*. Traverse City, MI: Myers Print Service.

Harrison, J. G., and M. A. Ogilvie. 1968. Immigrant mute swans in south-east England. *Wildfowl Trust Annual Report* 18: 85-87.

Hindman, L. J. 1982. Feral mute swan population status and problems in the Atlantic Flyway with special reference to Maryland's population. *Trumpeter Swan Society Conference* 8: 4-7.

Hindman, L. J., and R. L. Tjaden. 2014. Awareness and opinions of Maryland citizens toward Chesapeake Bay mute swans *Cygnus olor* and management alternative. *Wildfowl* 64: 167-185.

Hindman, L. J., and W. F. Harvey. 2004. Status and management of feral mute swans in Maryland. Pp. 11-17 in *Mute Swans and Their Chesapeake Bay Habitats: Proceedings of a Symposium* (M. C. Perry, ed.). US Geological Survey, Biological Resources Discipline Information and Technology Report USGS/BRD/ITR—2004-0005. 60 pp. . https://digitalcommons.unl.edu/usgspubs/138/

Hindman, L. J., R. A. Malecki, and C. M. Sousa. 2004. Mute swans in Maryland: Their status and a proposal for management (abstract). *Trumpeter Swan Society Conference* 19: 204.

Hindman, L. J., W. F. Harvey, and L. E. Conley. 2014. Spraying corn oil on mute swan *Cygnus olor* eggs to prevent hatching. *Wildfowl* 64: 186-196.

Huxley, J. S. 1947. Display of the mute swan. *British Birds* 40: 130-134.

Johnsgard, P. A., and J. Kear. 1968. A review of parental carrying of young by waterfowl. *Living Bird* 7: 89-102.

Johnston, W. 1935. Notes on the nesting of captive mute swans. *Wilson Bulletin* 47: 237-238.

Källander, H. 2005. Commensal association of waterfowl with feeding swans. *Waterbirds* 28: 326-330.

Kania, G. S., and H. R. Smith. 1986. Observations of agonistic interactions between a pair of feral mute swans and nesting waterfowl. *Connecticut Warbler* 6: 35-37.

Keane, E. M., and J. O'Halloran. 1992. The behavior of a wintering flock of mute swans *Cygnus olor* in southern Ireland. *Wildfowl* 43: 12-19.

Knapik, R. 2019. A research-based solution to Michigan's century-old mute swan problem. Paper presented at 25th Trumpeter Swan Society Conference, November 19-20, 2019, Alton, Illinois.

Knapton, R. W. 1993. Population status and reproductive biology of the mute swan, *Cygnus olor*, at Long Point, Lake Erie, Ontario. *Canadian Field-Naturalist* 77: 354-356.

Koechlein, A. 1971. Nest site selection by mute swans in the Grand Traverse Bay area, Michigan. MS thesis. East Lansing: Michigan State University.

Krull, J. N. 1970. Aquatic plant-macroinvertebrate association and waterfowl. *Journal of Wildlife Management* 34: 707–718.

Lever, C. 1987. *Naturalized Birds of the World*. Harlow, UK: Longman Scientific and Technical.

Lind, H. 1984. The rotation display of the mute swan *Cygnus olor*: Synchronised neighbour responses as instrument in the territorial defense strategy. *Ornis Scandinavia* 15: 98–104.

Lumsden, H. G. 1985. Foot clapping display of mute swans. *Bird Study* 32: 150–151.

Luukkonen, D. R. 2014. Mute swan population growth in Michigan in relation to management intensity. *Trumpeter Swan Society Conference* (abstract). 23 (unpaged).

Marshall, R. V. A. 1984. Alighting-display of mute swan. *British Birds* 77: 153–154.

Maryland Department of Natural Resources. 2001. *The Maryland Mute Swan Task Force Recommendations: A Summary of Information*. Annapolis: Maryland Department of Natural Resources. 55 pp. https://dnr.maryland.gov/wildlife/Documents/MuteSwan_TaskForceReport.pdf

Mathiasson, S. 1973. A moulting population of non-breeding mute swans with special reference to flight-feather moult, feeding ecology, and habitat selection. *Wildfowl* 24: 43–53.

———. 1981. Weight and growth rates of morphological characters of *Cygnus olor*. Pp. 379–389 in *Proceedings of the Second International Swan Symposium, Sapporo, 1980* (G. V. T. Matthews and M. Smart, eds.). Slimbridge, UK: International Waterfowl Research Bureau.

McCerry, R. H., C. Perrins, D. Wheeler, and S. Groves. 2002. Population structure, survival rates, and productivity of mute swans breeding in a colony in Abbotsbury, Dorset, England. Special Publication 1: Proceedings of the Fourth International Swan Symposium 2001 (E. C. Rees, S. L. Earnst, and J. Coulson, eds.). *Waterbirds* 25: 192–201.

Meyer, S. W., S. S. Badzinski, M. L. Schummer, and C. M. Sharp. 2012. Changes in summer abundance and distribution of mute swans along the lower Great Lakes of Ontario, 1986–2011. *Ontario Birds* 30: 48–63.

Michigan Department of Natural Resources. 2003. *Mute Swan Issues in Michigan* (history and status). Wildlife Issue Review Paper 12. 10 pp. https://www.michigan.gov/documents/dnr/mute_swan_issue_paper_June_30_2003_364890_7.pdf

Minton, C. D. T. 1968. Pairing and breeding of mute swans. *Wildfowl* 19: 41–60.

———. 1971. Mute swan flocks. *Wildfowl* 22: 71–88.

Munro, R. E., L. T. Smith, and J. J. Kupa. 1968. The genetic basis of color differences observed in the mute swan (*Cygnus olor*). *Auk* 85: 504–505.

Nelson, H. K. 1999. Mute swan populations, distribution, and management issues in the United States and Canada. *Trumpeter Swan Society Conference* 16: 125–132.

New York State Department of Environmental Conservation. 2013. *Management Plan for Mute Swans in New York State*. Albany: New York State Department of Environmental Conservation, Division of Fish, Wildlife, and Marine Resources.

Nummi, P., and L. Saari. 2003. Density-dependent decline of breeding success in an introduced, increasing mute swan *Cygnus olor* population. *Journal of Avian Biology* 34: 105–111.

O'Brien, M., and R. A. Askins. 1985. The effects of mute swans on native waterfowl. *Connecticut Warbler* 5: 27–31.

Ogilvie, M. A. 1967. Population changes and mortality of the mute swan in Britain. *Wildfowl Trust Annual Report* 18: 64–73.

O'Halloran, J., A. A. Myers, and P. F. Duggan. 1991. Lead poisoning in mute swans *Cygnus olor* in Ireland: A review. *Wildfowl* (Supplement No. 1): 389–395.

Owen, M., and C. J. Cadbury. 1975. The ecology and mortality of mute swans at the Ouse Washes, England. *Wildfowl* 25: 31–42.

Owens, S. 1995. Mute/whooper swan hybrid. *BBC Wildlife* 13(6): 45.

Perrins, C. M., and C. M. Reynolds. 1967. A preliminary study of the mortality of the mute swan, *Cygnus olor*. *Wildfowl Trust Annual Report* 18: 74–84.

Perrins, C. M., and J. Sears. 1991. Collisions with overhead wires as a cause of mortality in mute swan *Cygnus olor*. *Wildfowl* 42: 5–11.

Perrins, C. M., and M. A. Ogilvie. 1981. A study of the Abbotsbury mute swans. *Wildfowl* 32: 35–47.

Perrins, C. M., G. Cousquer, and J. Waine. 2003. A survey of blood lead levels in mute swans *Cygnus olor*. *Avian Pathology* 32: 205–212.

Perry, L. 2014. Mute swans, iconic and under attack! *The Osprey* 39(2) (March/April 2014): 1, 5.

Perry, M. C., P. C. Osenton, and E. J. R. Lohnes. 2004. Food habits of mute swans in the Chesapeake Bay. Pp. 31–35 in *Mute Swans and Their Chesapeake Bay Habitats: Proceedings of a Symposium* (M. C. Perry, ed.). US Geological Survey, Biological Resources Discipline Information and Technology Report USGS/BRD/ITR—2004-0005. 60 pp. . https://digitalcommons.unl.edu/usgspubs/138/

Petrie, S. A. 2004. Review of the status of mute swans on the Canadian side of the lower Great Lakes (abstract). *Trumpeter Swan Society Conference* 19: 201–202.

Petrie, S. A., and C. M. Francis. 2003. Rapid increase in the lower Great Lakes population of feral mute swans: A review and a recommendation. *Wildlife Society Bulletin* 31: 407–416.

Reese, J. G. 1975. Productivity and management of feral mute swans in Chesapeake Bay. *Journal of Wildlife Management* 39: 280–286.

———. 1980. Demography of European mute swans in Chesapeake Bay. *Auk* 97: 449–464.

Reichholf, V. J. 1984. On the function of territoriality in the mute swan (*Cygnus olor*). *Verhandlungen Ornithologischen Gesellschaft in Bayern* 24: 125–135. (In German, with English summary.)

Reiswig, B. 1986. Western mute swan population status and agency attitudes. *Trumpeter Swan Society Conference* 9: 116.

Reynolds, C. M. 1965. The survival of mute swan cygnets. *Bird Study* 12: 128–129.

Richman, A. 2009. Nutrient reserve dynamics of non-breeding mute swans (*Cygnus olor*) on the lower Great Lakes. Honors thesis. London: University of Western Ontario.

Scott, D. K. 1984a. Winter territoriality of mute swans *Cygnus olor*. *Ibis* 126: 168–176.

———. 1984b. Parent-offspring association in mute swans (*Cygnus olor*). *Zeitschrift für Tierpsychologie* 64: 74–86.

Scott, D. K., and M. E. Birkhead. 1983. Resources and reproductive performance in mute swans *Cygnus olor*. *Journal of Zoology* 200: 539–547.

Sears, J. 1989. Feeding activity and body condition of mute swans *Cygnus olor* in rural and urban areas of a lowland river system. *Wildfowl* 40: 88–98.

———. 1992. Extra-pair copulation by breeding male mute swans. *British Birds* 85: 558–559.

Sousa, C. M., R. A. Malecki, A. J. Lembo Jr., and L. J. Hindman. 2008. Monitoring habitat use by male mute swans in the Chesapeake Bay. *Proceedings of the Southeastern Association of Fish and Wildlife Agencies* 62: 88–93.

Stafford, J. D., M. W. Eichholz, and A. C. Phillips. 2012. Impacts of mute swans (*Cygnus olor*) on submerged aquatic vegetation in Illinois River Valley backwaters. *Wetlands* 32: 851–857.

Stone, W. B., and A. D. Masters. 1971. Aggression among captive mute swans. *New York Fish and Game Journal* 17: 50–52.

Swift, B. L., K. J. Clarke, R. A. Holevinski, and E. M. Cooper. 2013. *Status and Ecology of Mute Swans in New York State*. Albany: New York State Department of Environmental Conservation, Division of Fish, Wildlife, and Marine Resources. 37 pp. https://www.dec.ny.gov/docs/wildlife_pdf/muteswan-report.pdf

Tatu, K. S. 2006. An assessment of impacts of mute swans (*Cygnus olor*) on submerged aquatic vegetation (SAV) in Chesapeake Bay, Maryland. PhD diss. Morgantown: West Virginia University.

Tatu, K. S., J. T. Anderson, and L. J. Hindman. 2007. Predictive modeling for submerged aquatic vegetation (SAV) decline due to mute swans in the Chesapeake Bay. *Trumpeter Swan Society Conference* 20: 140–147.

Tatu, K. S., J. T. Anderson, L. J. Hindman, and G. Seidel. 2007a. Diurnal foraging activities of mute swans in Chesapeake Bay, Maryland. *Waterbirds* 30: 121–128.

———. 2007b. Mute swans' impact on submerged aquatic vegetation in Chesapeake Bay. *Journal of Wildlife Management* 71: 1431–1439.

Teale, C. L. 2011. A revised account of initial mute swan (*Cygnus olor*) introduction to the northeastern United States. *Biological Invasions* 13: 1729–1733.

Terenius, O. 2016. Windsurfing in mute swans (*Cygnus olor*). *Wilson Journal of Ornithology* 128: 628–631.

Therres, G. D., and D. F. Brinker. 2004. Mute swan interactions with other birds in the Chesapeake Bay. Pp. 43–46 in *Mute Swans and Their Chesapeake Bay Habitats: Proceedings of a Symposium* (M. C. Perry, ed.). US Geological Survey, Biological Resources Discipline Information and Technology Report USGS/BRD/ITR—2004-0005. 60 pp. . https://digitalcommons.unl.edu/usgspubs/138/

Wainwright, C. B. 1957. Results of wildfowl ringing at Abberton Reservoir, Essex, 1949–1966. *Wildfowl Trust Annual Report* 18: 28–35.

Walter, P. J., P. J. Bacon, and J Sears. 1991. An analysis of mute swan, *Cygnus olor*, breeding data. Pp. 151–156 in *Wildfowl: Supplement Number 1, Third IWRB International Swan Symposium, Oxford, England, December 9–13, 1989* (J. Sears and P. J. Bacon, eds.). Slimbridge, UK: International Waterfowl and Wetlands Research Bureau.

Watola, G. V., D. A. Stone, G. C. Smith, G. J. Forrester, A. E. Coleman, J. T. Coleman, M. J. Goulding, K. A. Robinson, and T. P. Milsom. 2003. Analyses of two mute swan populations and the effects of clutch reduction: Implications for population management. *Journal of Applied Ecology* 40: 565–579.

Weiloch, M. 1991. Population trends of the mute swan *Cygnus olor* in the Palearctic. *Wildfowl* (Special Issue 1): 22–32.

Willey, C. H. 1968a. The ecological significance of the mute swan in Rhode Island. Pp. 121–134 in *Transactions of the Northeast Section, the Wildlife Society*, vol. 25 (R. D. McDowell, ed.). Bedford, NH: Northeast Section, the Wildlife Society.

———. 1968b. The ecology, distribution, and abundance of the mute swan (*Cygnus olor*) in Rhode Island. MS thesis. Kingston: University of Rhode Island.

Willey, C. H., and B. F. Halla. 1972. *Mute Swans of Rhode Island*. Wildlife Pamphlet 8. West Kingston, RI: Division of Fish and Wildlife, Rhode Island Department of Natural Resources. 47 pp.

Williams, T. 1997. The ugly swan. *Audubon* 97(6): 26–32.

Włodarczyk, R., M. Wieloch, S. Czyż, P. T. Dolata, and P. Minias. 2013. Natal and breeding dispersal in mute swans *Cygnus olor*: Influence of sex, mate switching, and reproductive success. *Acta Ornithologica* 48: 237–244.

Wood, K. A., M. J. Brown, R. L. Cromie, G. M. Hilton, C. Mackenzie, J. L. Newth, D. J. Pain, C. M. Perrins, and E. C. Rees. 2019. Regulation of lead fishing weights results in mute swan population recovery. *Biological Conservation* 230: 67–74.

Wood, R., and W. L. Gelston. 1972. *Preliminary Report: The Mute Swans of Michigan's Grand Traverse Bay Region*. Report 2683. Lansing: Wildlife Division, Michigan Department of Natural Resources.

Trumpeter Swan

Abel, R. 1993. Trumpeter swan reintroduction. MS thesis. Madison: University of Wisconsin.

American Ornithologists' Union (AOU). 1998. *Check-list of North American Birds*. 7th ed. Washington, DC: American Ornithologists' Union. (Includes annual supplements to 2019.)

Anderson, D. R., R. C. Herron, and B. Reiswig. 1986. Estimates of annual survival rates of trumpeter swans banded 1949–1982 at Red Rock Lakes National Wildlife Refuge, Montana. *Journal of Wildlife Management* 50: 218–221.

Anderson, P. S. 1992. Changing land use and trumpeter swans in the Skagit Valley. *Trumpeter Swan Society Conference* 13: 150–156.

————. 1993. Distribution and habitat selection by wintering trumpeter swans *Cygnus buccinator* in the Skagit Valley, Washington. MS thesis. Seattle: University of Washington.

————. 1994. Distribution and habitat selection by wintering trumpeter swans in the lower Skagit Valley, Washington. *Trumpeter Swan Society Conference* 14: 61-71.

————. 2004. The Pacific Coast Population—historical perspective and future concerns. *Trumpeter Swan Society Conference* 19: 3-8.

Andrews, R., and D. Hoffman. 2007. Iowa's trumpeter swan restoration program—a 2005 update. *Trumpeter Swan Society Conference* 20: 3-10.

Anglin, R. M. 1999. Rocky Mountain Population of trumpeter swans: The winter range expansion program. *Trumpeter Swan Society Conference* 16: 56-58.

Babineau, F. M. 2004. Winter ecology of trumpeter swans in southern Illinois. MS thesis. Carbondale: Southern Illinois University.

Babineau, F. M., and D. Holm. 2004. Winter distribution and habitat use of trumpeter swans in Illinois. *Trumpeter Swan Society Conference* 19: 166-174.

Bailey, T. N., E. E. Bangs, and M. F. Portner. 1986. Trumpeter swan surveys and studies on the Kenai National Wildlife Refuge and Kenai Peninsula, Alaska, 1957-1984 (abstract). *Trumpeter Swan Society Conference* 9: 64.

Bailey, T. N., E. E. Bangs, and V. Bailey. 1980. Back carrying of young by trumpeter swans. *Wilson Bulletin* 92: 413.

Bailey, T. N., M. F. Portner, E. E. Bangs, W. W. Larned, R. A. Richey, and R. L. Delaney. 1990. Summer and migratory movements of trumpeter swans using the Kenai National Wildlife Refuge, Alaska. *Trumpeter Swan Society Conference* 11: 72-81.

Bales, B. 1992. Update on the Pacific Coast Population Swan Management Plan. *Trumpeter Swan Society Conference* 13: 131-132.

Bales, B. D., and D. Kraege. 1992. Management challenges related to Pacific Coast Population trumpeter swans in Oregon and Washington. *Trumpeter Swan Society Conference* 13: 157-159.

————. 1994. Management challenges in the 1990s related to Pacific Coast Population trumpeter swans in Oregon and Washington. *Trumpeter Swan Society Conference* 14: 98-100.

Ball, I. J., E. O. Garton, and R. E. Shea. 2001. History, ecology, and management of the Rocky Mountain Population of trumpeter swans: Implications for restoration. *Trumpeter Swan Society Conference* 17: 45-49.

Banko, W. 1960. *The Trumpeter Swan: Its History, Habits, and Population in the United States.* Washington, DC: US Department of the Interior, Fish and Wildlife Service, North American Fauna No. 63. 214 pp.

Banko, W. E., and A. W. Schorger. 1976. Trumpeter swan. Pp. 5-71 in *Handbook of North American Birds*, vol. 2.; Waterfowl, pt. 1 (R. S. Palmer, ed.). New Haven, CT: Yale University Press.

Barrett, V. A., and E. R. Vyse. 1982. Comparative genetics of three trumpeter swan populations. *Auk* 99: 103-108.

Bart, J., C. D. Mitchell, M. N. Fisher, and J. A. Dubvsky. 2007. Detection ratios on winter surveys of Rocky Mountain trumpeter swans *Cygnus buccinator*. *Wildfowl* 57: 21-28.

Bartok, N. D. 2004. 2002 nesting success of the trumpeter swan (*Cygnus buccinator*) population that frequents the Wye Marsh, Ontario. *Trumpeter Swan Society Conference* 19: 159-165.

Bauer, R. D. 1991. US Fish and Wildlife Service involvement in winter management of the Rocky Mountain Population of trumpeter swans. *Trumpeter Swan Society Conference* 12: 195.

Beaulieu, R. 1992. Saskatchewan trumpeter swans—1991. *Trumpeter Swan Society Newsletter* 21(1): 7.

Becker, D. M. 2001. Trumpeter swan reintroduction on the Flathead Indian Reservation. *Trumpeter Swan Society Conference* 17: 103-106.

————. 2019. Restoration of trumpeter swans on the Flathead Indian Reservation and Northwestern Montana and Tribal public outreach. Paper presented at 25th Trumpeter Swan Society Conference, November 19-20, 2019, Alton, Illinois.

Becker, D. M., and J. S. Lichtenberg. 2004. Trumpeter swan reintroduction on the Flathead Indian Reservation—an overview and update. *Trumpeter Swan Society Conference* 19: 128–133.

Becker, D. M. and J. S. Lichtenberg. 2007. Trumpeter swan reintroduction on the Flathead Indian Reservation. *Trumpeter Swan Society Conference* 20: 100–105.

Berquist, J. 1990. Status report on Turnbull National Wildlife Refuge's trumpeter swan population. *Trumpeter Swan Society Conference* 11: 117.

Beyersbergen, G. W., ed. 2007. *The 2005 International Trumpeter Swan Survey in Alberta, Saskatchewan, Manitoba, and the Northwest Territories.* Edmonton, AB: Canadian Wildlife Service Technical Report Series, 485: 1–45.

Beyersbergen, G. W., and L. Shandruk. 1994. Interior Canada subpopulation of trumpeter swans—status 1992. *Trumpeter Swan Society Conference* 14: 103–110.

Beyersbergen, G. W., and R. Kaye. 2007. Elk Island National Park trumpeter swan reintroduction—2005 update. *Trumpeter Swan Society Conference* 20: 88–97.

Beyersbergen, G. W., M. Heckbert, R. Kaye, T. Sallows, and P. Latour. 2007. The 2005 international survey of trumpeter swans in Alberta, Saskatchewan, Manitoba, and the Northwest Territories. *Trumpeter Swan Society Conference* 20: 78–87.

Bishop, A., S. Comeau, J. Dubovsky, and A. Araya. 2011. Landscape-level habitat use by trumpeter swans in the Sandhills of Nebraska and South Dakota (abstract). *Trumpeter Swan Society Conference* 22 (unpaged).

Blus, L. J., R. Stroud, B. Reiswig, and T. McEneaney. 1990. Lead poisoning and other mortality factors of trumpeter swans (abstract). *Trumpeter Swan Society Conference* 11: 152.

Bogdan, L. 2004. Development of a detailed landscape plan to support overwintering and migrating waterfowl for the Fraser River delta (abstract). *Trumpeter Swan Society Conference* 19: 15.

Bollinger, K. S., and R. King. 2002. Activity budgets of nesting trumpeter swans in interior Alaska. Special Publication 1: Proceedings of the Fourth International Swan Symposium 2001 (E. C. Rees, S. L. Earnst, and J. Coulson, eds.). *Waterbirds* 25: 285–292.

Bouffard, S. H. 1986a. Rocky Mountain Population (Tristate flock): Status of trumpeter swans at Camas National Wildlife Refuge 1983–1984. *Trumpeter Swan Society Conference* 9: 54–55.

———. 1986b. Pacific Coast Population: Status of trumpeter swan restoration flocks, 1983–84. *Trumpeter Swan Society Conference* 9: 89–91.

———. 2000. Recent changes in winter distribution of RMP [Rocky Mountain Population] trumpeter swans. *Trumpeter Swan Society Conference* 17: 53–59.

Boyd, W. S., and A. Breault. 2004. Trumpeter swans wintering in southwest British Columbia: An assessment of status and trends (abstract). *Trumpeter Swan Society Conference* 19: 10.

Breault, A. M. 2008. Breeding distribution and abundance of trumpeter swans (*Cygnus buccinator*) in British Columbia, summer 2015 (abstract). *Trumpeter Swan Society Conference* 21 (unpaged).

Breault, A., M. Campbell, and G. Raven. 2016. Delivery of the 2015 continental trumpeter swan survey in western Canada (abstract). *Trumpeter Swan Society Conference* 24 (unpaged).

Brininger, W. L., Jr., and L. C. Deede. 2011. The status of trumpeter swans at Tamarac National Wildlife Refuge: A quarter century after reintroduction (abstract). *Trumpeter Swan Society Conference* 22 (unpaged).

Brooks, A. 1926. The present status of the trumpeter swan. *Condor* 28: 129.

Brown, C. S., and J. Luebbert. 2000. Field triage and rehabilitation of swans. *Trumpeter Swan Society Conference* 17: 170–76.

Brown, S. 1990. A status report of the introduced trumpeter swan population at Ruby Lakes National Wildlife Refuge, Nevada. *Trumpeter Swan Society Conference* 11: 123–124.

Buffet, D., and M. R. Petrie. 2004. Spatial and temporal use of estuary and upland habitat by waterfowl wintering on the Fraser River delta and North Puget Sound (abstract). *Trumpeter Swan Society Conference* 19: 15.

Burgess, H. H. 1986. Potential trumpeter swan restoration. *Trumpeter Swan Society Conference* 9: 97–111.

———. 1997. Significant observations of trumpeter swans in Saskatchewan. *Blue Jay* 55: 35–40.

———. 2001a. History of the High Plains trumpeter swan restoration. *North American Swans* 30: 6–14.

———. 2001b. North Dakota trumpeter swan observations. *North American Swans* 30: 21–24.

———. 2002a. Trumpeter swan myths, movements, and migrations of the High Plains Flock. *North American Swans* 31: 7–11.

———. 2002b. High Plains trumpeter swan nesting ecology. *North American Swans* 31(1): 5–6.

Burgess, H. H., and M. E. Bote. 1999. Observations of trumpeter swans in Manitoba. *North American Swans* 28: 25–30.

Burgess, H. H., and R. Burgess. 1988. Elk Island National Park trumpeter swan restoration experimental project. *Trumpeter Swan Society Conference* 10: 78–88.

———. 1991. History of trumpeter swan restoration to the Upper Midwest. *Trumpeter Swan Society Conference* 12: 131–132.

———. 1997. Trumpeter swans once wintered in Texas—why not now? *North American Swans* 26(2): 50–53.

———. 1998. The Nebraska trumpeter swans. *North American Swans* 27(1): 30–31.

Burgess, H. H., R. Burgess, and D. K. Weaver. 1990. Potential trumpeter swan restoration and expansion. *Trumpeter Swan Society Conference* 11: 62–64.

Burgess, H. H., R. Burgess, and M. Bote. 1999a. Trumpeter swans once wintered on the lower Mississippi River. Why not now? *Trumpeter Swan Society Conference* 16: 3–5.

———. 1999b. Developing trumpeter swan wintering areas. *Trumpeter Swan Society Conference* 16: 25–26.

Bush, D. 2019. Incubation behavior and genetic attributes of trumpeter swans restored to Grays Lake National Wildlife Refuge. Paper presented at 25th Trumpeter Swan Society Conference, November 19–20, 2019, Alton, Illinois.

Caithamer, D. F. 1996. 1995 Survey of Trumpeter Swans in North America. Unpublished report. Laurel, MD: US Fish and Wildlife Service, Division of Migratory Bird Management.

———. 2001. Trumpeter Swan Population Status. 2000. Unpublished report. Laurel, MD: US Fish and Wildlife Service, Division of Migratory Bird Management.

Canniff, R. S. 1986. Wintering trumpeter swans, Skagit Valley, Washington: Update 1980–1984. *Trumpeter Swan Society Conference* 9: 71–75.

Carrick, W. H. 1991. Use of imprinted swans to establish a migratory population. *Trumpeter Swan Society Conference* 12: 143.

———. 1999. Induced migration using ultralite aircraft. *Trumpeter Swan Society Conference* 16: 115–116.

Carroll, D., and B.L. Swift. 2000. Status of the trumpeter swan in New York State. *Kingbird* 50: 232–236.

Childress, D. 1986. Trumpeter swan expansion in Montana. *Trumpeter Swan Society Conference* 9: 47.

Churchill, B. P. 1988. Potential trumpeter swan nesting habitat in northeastern British Columbia. *Trumpeter Swan Society Conference* 10: 29–35.

Coale, H. K. 1915. The present status of the trumpeter swan. *Auk* 32: 82–90.

Cochran, D. K. 1970. Food preferences in captivity and weight gains in trumpeter swans. *Trumpeter Swan Society Newsletter* 4: 9–13.

Comeau-Kingfisher, S., and M. Vrtiska. 2007. Fall Trumpeter Swan Survey of the High Plains Flock. Unpublished report. Martin SD: US Fish and Wildlife Service, Lacreek National Wildlife Refuge.

Comeau-Kingfisher, S., and T. Koerner. 2005. *Management Plan for the High Plains Trumpeter Swan Flock*. Martin, SD: US Fish and Wildlife Service, Lacreek National Wildlife Refuge.

———. 2007. Status of the High Plains flock of trumpeter swans in 2005. *Trumpeter Swan Society Conference* 20: 23–27.

Compton, D. 1985. Captive trumpeter swan survey results. *Trumpeter Swan Society Conference* 10: 146–148.

———. 1991a. Results of the 1988 captive trumpeter swan survey. *Trumpeter Swan Society Conference* 12: 45.

———. 1991b. Trumpeter swan banding protocol—a survey of the banders. *Trumpeter Swan Society Conference* 12: 49–52.

———. 1991c. Hennepin Park's trumpeter swan restoration update. *Trumpeter Swan Society Conference* 12: 91–94.

———. 1991d. Transport of trumpeter swan eggs and cygnets. *Trumpeter Swan Society Conference* 12: 147.

Compton, D. C. 1996. Interior Population status report, highlights, and trends, December 1994. *Trumpeter Swan Society Conference* 15: 18–37.

Conant, B. 1991. Alaskan trumpeter swan status report (Abstract). *Trumpeter Swan Society Conference* 12: 9.

Conant, B., J. I. Hodges, D. J. Groves, and J. G. King. 1991. *Alaska Trumpeter Swan Status Report.* Juneau, AK: US Fish and Wildlife Service.

———. 1992. The 1990 census of trumpeter swans on Alaskan nesting habitats. *Trumpeter Swan Society Conference* 13: 133–146.

———. 1994. A potential summer population of trumpeter swans (*Cygnus buccinator*) for Alaska (abstract). *Trumpeter Swan Society Conference* 14: 5-6.

———. 1996. *An Atlas of the Distribution of Trumpeter Swans in Alaska and Instructions for the Use of an Archival System.* Juneau, AK: US Fish and Wildlife Service.

———. 1999. The 1995 census of trumpeter swans on Alaskan nesting habitats. *Trumpeter Swan Society Conference* 16: 75–97.

———. 2002. Census of trumpeter swans on Alaskan nesting habitats, 1968–2000. Special Publication 1: Proceedings of the Fourth International Swan Symposium 2001 (E. C. Rees, S. L. Earnst, and J. Coulson, eds.). *Waterbirds* 25: 3-7.

———. 2007. *Alaska Trumpeter Swan Status Report—2005.* Juneau, AK: US Fish and Wildlife Service, Migratory Bird Management.

———. 2007. The 2005 census of trumpeter swans on Alaskan nesting habitats (abstract). *Trumpeter Swan Society Conference* 20: 107-112.

Conant, B., J. I. Hodges, J. G. King, and A. Loranger. 1986. Alaska trumpeter swan status report—1984. *Trumpeter Swan Society Conference* 9:76–89.

Conant, B., J. I. Hodges, J. G. King, and S. L. Cain. 1988. Alaska trumpeter swan status report—1985. *Trumpeter Swan Society Conference* 10: 121–129.

Cooper, B. A., and R. J. Ritchie. 1990. Migration of trumpeter and tundra swans in east-central Alaska during spring and fall, 1987. *Trumpeter Swan Society Conference* 11: 82–91.

Cooper, B., J. King, and R. J. Ritchie. 1991. Swan migration routes in the Nelchina Basin, Alaska, during spring migration 1989 (abstract). *Trumpeter Swan Society Conference* 12:11.

Cooper, J. A. 1979. Trumpeter swan nesting behaviour. *Wildfowl* 30: 55–71.

Cooper, J. A., and D. K. Weaver, eds. 1986. Trumpeter Swan Bibliography. (Available from the Trumpeter Swan Society, 12615 Rockford Road, Plymouth, MN 55441.)

Corace, R. G., III, D. L. McCormick, and V. Cavalieri. 2006. Population growth parameters of a reintroduced trumpeter swan flock, Seney National Wildlife Refuge, Michigan, USA (1991-2004). *Waterbirds* 29: 38-42.

Cornely, J. E., S. A. Petrie, and L. J. Hindman. 2014. Status of swans breeding in North America (abstract). *Trumpeter Swan Society Conference* 23 (unpaged).

Cornely, J. E., S. P. Thompson, E. McLaury, and L. D. Napier. 1985. A summary of trumpeter swan production at Malheur National Wildlife Refuge, Oregon. *Murrelet* 66: 50–55.

Czarnowski, K. 1986. Yellowstone National Park policy for managing trumpeter swans. *Trumpeter Swan Society Conference* 9: 27-28.

Dalgleish, J. J. 1880. List of occurrences of North American birds in Europe. *Bulletin of the Nuttall Ornithological Club* 5(4): 210-221.

Degernes, L. A. 1991. The Minnesota trumpeter swan lead poisoning crisis of 1988-89. *Trumpeter Swan Society Conference* 12: 114-118.

Degernes, L. A., and P. T. Redig. 1990. Diagnosis and treatment of lead poisoning in trumpeter swans. *Trumpeter Swan Society Conference* 11: 153-158.

Degernes, L. A., and R. K. Frank. 1991. Minnesota trumpeter swan mortality, January 1988-June 1989. *Trumpeter Swan Society Conference* 12: 111-113.

Degernes, L. A., P. T. Redig, and M. Freeman. 1991. New treatments for lead poisoned trumpeter swans. *Trumpeter Swan Society Conference* 12: 161-162.

Degernes, L., S. Heilman, M. Trogdon, M. Jordan, M. Davison, D. Kraege, M. Correa, and P. Cowen. 2006. Epidemiologic investigation of lead poisoning in trumpeter and tundra swans in Washington State, USA, 2000-2002. *Journal of Wildlife Diseases* 42(2): 345-358.

Degernes, L., S. Kittelson, M. Linck, and P. Manthey. 2002. Survival and reproductive success of trumpeter swans after treatment for lead poisoning. Special Publication 1: Proceedings of the Fourth International Swan Symposium 2001 (E. C. Rees, S. L. Earnst, and J. Coulson, eds.). *Waterbirds* 25: 368-374.

Dennington, M. 1988. Trumpeter swan habitat in southern Yukon. *Trumpeter Swan Society Conference* 10: 36-41.

Denson, E. P. 1970. The trumpeter swan, *Olor buccinator*: A conservation success and its lessons. Biological Conservation 2: 253-256.

De Vos, A. 1964. Observations on the behaviour of captive trumpeter swans during the breeding season. *Ardea* 52: 166-189.

Doyle, T. J. 1994. Expansion of trumpeter swans in the upper Tanana Valley, Alaska. *Trumpeter Swan Society Conference* 14: 7-18.

Drewien, R. C., and D. S. Benning. 1997. Status of tundra swans and trumpeter swans in Mexico. *Wilson Bulletin* 109: 693-701.

Drewien, R. C., and S. H. Bouffard. 1994. Winter body mass and measurements of trumpeter swans *Cygnus buccinator*. *Wildfowl* 45: 22-32. https://wildfowl.wwt.org.uk/index.php/wildfowl/article/view/936

Drewien, R. C., J. T. Herbert, T. W. Aldrich, and S. H. Bouffard. 1999. Detecting trumpeter swans harvested in tundra swan hunts. *Wildlife Society Bulletin* 27: 95-102.

Drewien, R. C., K. Clegg, and R. E. Shea. 2002. Use of winter translocations to expand distribution of trumpeter swans in the western United States. Special Publication 1: Proceedings of the Fourth International Swan Symposium 2001 (E. C. Rees, S. L. Earnst, and J. Coulson, eds.). *Waterbirds* 25: 138-142.

Drewien, R. C., K. R. Clegg, and M. N. Fisher. 1992. Winter capture of trumpeter swans at Harriman State Park, Idaho, and Red Rock Lakes National Wildlife Refuge, Montana. *Trumpeter Swan Society Conference* 13: 38-46.

Drewien, R. C., R. E. Shea, B. Conant, J. S. Hawkings, and N. Hughes. 2004. Satellite tracking trumpeter swans from the Yukon Territory (abstract). *Trumpeter Swan Society Conference* 19: 43.

Dubovsky, J. A. 2008. Status of Rocky Mountain Population trumpeter swans, 2007 (abstract). *Trumpeter Swan Society Conference* 21 (unpaged).

Dubovsky, J. A., and J. E. Cornely. 2002. *An Assessment of Information Pertaining to the Status of Trumpeter Swans* (Cygnus buccinator). Denver, CO: US Fish and Wildlife Service.

Ducey, J. E. 1999. History and status of the trumpeter swan in the Nebraska Sand Hills. *North American Swans* 28: 31-39.

Eaton, J. 1986. Trumpeter swans at Harriman State Park. *Trumpeter Swan Society Conference* 9: 51-52.

Eichholz, M. W., and D. M. Varner. 2007. Survival of Wisconsin Interior Population of trumpeter swans. *Trumpeter Swan Society Conference* 20: 45-52.

Eichholz, M. W., and D. Varner. 2008. Winter ecology of trumpeter swans in southern Illinois. Final Report, Federal Aid Project W-142-R-(3-5). Carbondale: Southern Illinois University. 10 pp. http://opensiuc.lib.siu.edu/cwrl_fr/2

Engelhardt, K. A. M., J. A. Kadlec, T. W. Aldrich, and V. L. Roy. 1999. The Utah trumpeter swan reintro-duction program: Proposal to evaluate reintroduction success. *Trumpeter Swan Society Conference* 16: 61–65.

Engelhardt, K. A. M., J. A. Kadlec, V. L. Roy, and J. A. Powell. 2000. Evaluation of translocation criteria: Case study with trumpeter swans (*Cygnus buccinator*). *Biological Conservation* 94: 173–181.

Eyraud, E. 1986. Harriman State Park: Background, management, and trumpeter swans. *Trumpeter Swan Society Conference* 9: 48–51.

Farley, L. C. 1980. The behavioral ecology of trumpeter swans wintering in southeast Alaska. MS thesis. Pocatello: Idaho State University.

Fjetland, C. A. 1974. Trumpeter swan management in the National Wildlife Refuge system. *Transactions of the North American Wildlife and Natural Resources Conference* 39: 136–141.

Fowler, G. M. 1999. Trumpeter swans in the community—Comox Valley, British Columbia. *Trumpeter Swan Society Conference* 16: 98–99.

Fowler, G. M., and B. Wareham. 1996. Comox Valley Waterfowl Management Project 1991–94 Report: A report on trumpeter swan management in the Comox Valley, British Columbia. *Trumpeter Swan Society Conference* 15: 44–47.

Froelich, A. J., J. C. Johnson, and D. M. Lodge. 1999. Food preferences of mute and trumpeter swans. *Trumpeter Swan Society Conference* 16: 133.

Gale, R. S. 1988. Trumpeter swan winter habitat relationships in the Tri-state area. *Trumpeter Swan Society Conference* 10: 54–56.

———. 1990a. Status of trumpeter swans in Idaho. *Trumpeter Swan Society Conference* 11: 4–5.

———. 1990b. Results of the cooperative Rocky Mountain Population trumpeter swan study. *Trumpeter Swan Society Conference* 11: 34–37.

Gale, R. S., E. O. Garton, and I. J. Ball. 1987. *The History, Ecology, and Management of the Rocky Mountain Population of Trumpeter Swans*. Missoula: Montana Cooperative Wildlife Research Unit, US Fish and Wildlife Service.

Gillette, L. N. 1988. Status report for the Hennepin Parks' trumpeter swan restoration project. *Trumpeter Swan Society Conference* 10: 104–108.

———. 1990a. The impact of nesting trumpeter swans on other species of waterfowl. *Trumpeter Swan Society Conference* 11: 162–163.

———. 1990b. Causes of mortality for trumpeter swans in central Minnesota, 1980–1987. *Trumpeter Swan Society Conference* 11: 148–151.

———. 1991a. The Trumpeter Swan Society draft position paper on tundra swan hunting. *Trumpeter Swan Society Conference* 12: 59–62.

———. 1991b. Ways to reduce the potential for lead poisoning in trumpeter swans. *Trumpeter Swan Society Conference* 12: 119–121.

———. 1991c. Options for establishing migratory populations of Interior Population trumpeter swans. *Trumpeter Swan Society Conference* 12: 136–138.

———. 1991d. Need for a coordinated restoration approach for the Interior Population of trumpeter swans. *Trumpeter Swan Society Conference* 12: 133.

———. 1996. Building a migratory tradition for the Interior Population of trumpeter swans. *Trumpeter Swan Society Conference* 15: 99–103.

———. 1999. Why is it so hard to establish a migratory population of trumpeter swans? *Trumpeter Swan Society Conference* 16: 21–24.

———. 2000a. What needs to be done to complete the restoration of the Interior Population of trumpeter swans? *Trumpeter Swan Society Conference* 17: 35–38.

———. 2000b. Perspectives of the Trumpeter Swan Society on management of the Rocky Mountain Population of trumpeter swans. *Trumpeter Swan Society Conference* 17: 82–84.

———. 2007. Is migration necessary for restoration of trumpeter swans in the Midwest? *Trumpeter Swan Society Conference* 20: 55–57.

Gillette, L. N., and M. H. Linck. 2004. Population status and management options for the Interior Population of trumpeter swans. *Trumpeter Swan Society Conference* 19: 139–147.

Gillette, L. N., and R. Shea. 1995. An evaluation of trumpeter swan management today and a vision for the future. *Transactions of the North American Wildlife and Natural Resources Conference* 60: 258–265.

Gomez, D. 2001. 1999 fall survey of the Rocky Mountain Population (RMP) of trumpeter swans, US flocks. *Trumpeter Swan Society Conference* 17: 50–52.

Gomez, D., and E. Scheuering. 1996. Termination of artificial feeding at Red Rock Lakes National Wildlife Refuge, Montana. *Trumpeter Swan Society Conference* 15: 62–69.

Grant, T. A. 1991. Foraging ecology of trumpeter swans breeding on the Copper River delta, Alaska. MS thesis. St. Paul: University of Minnesota.

Grant, T., and P. Henson. 1991. Habitat use by trumpeter swans breeding on the Copper River, Alaska (abstract). *Trumpeter Swan Society Conference* 12: 171.

Grant, T. A., P. Henson, and J. A. Cooper. 1994. Feeding ecology of trumpeter swans breeding in south central Alaska. *Journal of Wildlife Management* 58: 774–780.

———. 1997. Feeding behaviour of trumpeter swans *Cygnus buccinator*. *Wildfowl* 48: 6–15.

Griswold, J. A. 1965. We raise the first trumpeter swans. *America's First Zoo* 17: 17–20.

Groves, D. J., compiler. 2012. *The 2010 North American Trumpeter Swan Survey: A Cooperative North American Survey*. Juneau, AK: US Fish and Wildlife Service. 17 pp.

———. 2017. *The 2015 North American Trumpeter Swan Survey*. 26 pp. https://www.fws.gov/migratorybirds/pdf/surveys-and-data/NATrumpeterSwanSurvey_2015.pdf

Groves, D. J., B. Conant, and J. I. Hodges. 1997. A summary of Alaska trumpeter swan surveys 1996. *North American Swans* 26(2): 45–49.

Groves, D. J., B. Conant, E. Mallek, and D. Logan. 2002. Trumpeter swan surveys on the Chugach National Forest 2001—an update. *North American Swans* 31(1): 19–20.

Groves, D. J., B. Conant, J. Sarvis, and D. Logan. 2001. Trumpeter swan surveys on the Chugach National Forest 2000—an update. *North American Swans* 30(1): 51–54.

Groves, D. J., B. Conant, R. J. King, and D. Logan. 1998. Trumpeter swan surveys on the Chugach National Forest 1997. *North American Swans* 27(1): 36–45.

Groves, D. J., B. Conant, W. W. Larned, and D. Logan. 1999. Trumpeter swan surveys on the Chugach National Forest 1998—an update. *North American Swans* 28(1): 16–21.

Hamer, C. A. 1990. Winter behavior of trumpeter swans in northwest Washington. MS thesis. Bellingham: Western Washington University.

Hammer, D. A. 1970. Trumpeter swan carrying young. *Wilson Bulletin* 82: 324–325.

Hammer, D. (moderator). 1986. Panel discussion: Coordinating management of the Rocky Mountain trumpeter swan population and the role of the Trumpeter Swan Society. *Trumpeter Swan Society Conference* 9: 55–61.

Hampton, P. D. 1981. The wintering and nesting behavior of the trumpeter swan. MS thesis. Missoula: University of Montana.

Hanauska-Brown, L. A. 2004. Southeastern Idaho trumpeter swan translocations and observations 2001–2003—project update. *Trumpeter Swan Society Conference* 19: 101–107.

Handrigan, S. 2013. Evaluation of breeding range expansion of trumpeter swans (*Cygnus buccinator*) re-introduced in southwestern Ontario. Honor's thesis. London: University of Western Ontario.

Handrigan, S. A., M. L. Schummer, S. A. Petrie, and D. R. Norris. 2016. Range expansion and migration of trumpeter swans *Cygnus buccinator* re-introduced in southwest and central Ontario. *Wildfowl* 66: 60–74.

Hansen, H. A. 1973. Trumpeter swan management. *Wildfowl* 24: 27–32.

Hansen, H. A., P. E. K. Shepherd, J. G. King, and W. A Troyer. 1971. *The Trumpeter Swan in Alaska*. Wildlife Monographs No. 26. 83 pp.

Hansen, J. L. 1991. Iowa's role in trumpeter swan restoration. *Trumpeter Swan Society Conference* 12: 141.

Harwood, C. M. 2010. Trumpeter swan survey, Kanuti National Wildlife Refuge, August–September 2010. Unpublished progress report. Fairbanks, AK: US Fish and Wildlife Service, Kanuti National Wildlife Refuge.

Hawkings, J. S. 1990. Spring staging areas for trumpeter swans in the Southern Lakes Region of Yukon (abstract). *Trumpeter Swan Society Conference* 11: 33.

———. 2000. Design and effectiveness of the 1995 Yukon/Northern British Columbia trumpeter swan survey: An appropriate technique for 2000 and beyond? *Trumpeter Swan Society Conference* 17: 145–153.

———. 2007. The Yukon and northwestern British Columbia trumpeter swan survey, 2005. *Trumpeter Swan Society Conference* 20: 61–78.

Hawkings, J. S., A. Breault, S. Boyd, M. Norton, G. Beyersbergen, and P. Latour. 2002. Trumpeter swan numbers and distribution in western Canada, 1970–2000. Special Publication 1: Proceedings of the Fourth International Swan Symposium 2001 (E. C. Rees, S. L. Earnst, and J. Coulson, eds.). *Waterbirds* 25: 8–21.

Hawkings, J. S., and N. L. Hughes. 1994. Recruitment and overwinter survival of Pacific Coast trumpeter swans as determined from age ratio counts. *Trumpeter Swan Society Conference* 14: 37–47.

Hemker, T. P. 2004. The trumpeter swan implementation plan—an overview. *Trumpeter Swan Society Conference* 19: 134–135.

Henderson, C. 2016. History of trumpeter swan restoration in Minnesota (abstract). *Trumpeter Swan Society Conference* 24 (unpaged).

Henson, P. 1991. The nocturnal behavior of breeding trumpeter swans. *Trumpeter Swan Society Newsletter* 20(2): 11–12.

Henson, P., and J. A. Cooper. 1992. Division of labour in breeding trumpeter swans *Cygnus buccinator*. *Wildfowl* 43: 40–48.

———. 1993. Trumpeter swan incubation areas of differing food quality. *Journal of Wildlife Management* 57: 709–716.

Henson, P., and T. A. Grant. 1991. The effects of human disturbance on trumpeter swan breeding behavior. *Wildlife Society Bulletin* 19: 248–257.

Herbert, J. 1992. Rocky Mountain Population of trumpeter swans—a Pacific Flyway Study Committee perspective. *Trumpeter Swan Society Conference* 13: 19–21.

———. 1994. Pacific Flyway experimental "general" swan hunting season—a proposal. *Trumpeter Swan Society Conference* 14: 133–136.

Herwig, C. 2019. Update from MN trumpeter swan reintroduction in Minnesota. Paper presented at 25th Trumpeter Swan Society Conference, November 19–20, 2019, Alton, Illinois.

Hills, L. 2004. Spring and fall migration and pond usage by trumpeter swans, Cochrane area, Alberta, Canada, 2002. *Trumpeter Swan Society Conference* 19: 60–70.

Hinds, D. S., and W. A. Calder. 1971. Tracheal dead space in respiration of birds. *Evolution* 25: 429–440.

Hines, M. E. 1991a. Minnesota DNR trumpeter swan restoration efforts—1989 status report. *Trumpeter Swan Society Conference* 12: 97–99.

———. 1991b. Minnesota DNR efforts—the selection of wetlands for release of 2-year-old trumpeter swans. *Trumpeter Swan Society Conference* 12: 100–104.

Hodges, J. I., B. Conant, and S. L. Cain. 1988. Alaska trumpeter swan 1986 sample, and recommendations for a continent-wide sampling scheme (abstract). *Trumpeter Swan Society Conference* 10: 130.

———. A summary of the 1987 Alaska trumpeter swan surveys. *Trumpeter Swan Society Conference* 11: 68–71.

Hoffman, D. 2016. Status of trumpeter swan restoration and the promotion of wetlands in Iowa (abstract). *Trumpeter Swan Society Conference* 24 (unpaged).

———. 2019. Status of trumpeter swan restoration and the promotion of wetlands in Iowa. Paper presented at 25th Trumpeter Swan Society Conference, November 19–20, 2019, Alton, Illinois.

Holbek, N. 1994. Tools for dealing with land use problems on the coastal wintering areas, the agricultural land reserve. *Trumpeter Swan Society Conference* 14: 89–92.

Holton, G. 1982. Habitat use by trumpeter swans in the Grande Prairie region of Alberta. MS thesis. Calgary, AB: University of Calgary.

———. 1998. An overview of trumpeter swans in the Grande Prairie region, 1957–1986. *Trumpeter Swan Society Conference* 10: 11–17.

Howie, R. R. 1994. Trumpeter swans wintering in the Thompson-Okanagan areas of British Columbia. *Trumpeter Swan Society Conference* 14: 49–60.

Houston, C. S., M. I. Houston, and H. M. Reeves. 1997. The 19th-century trade in swan skins and quills. *Blue Jay* 55(1): 24–34.

Hughlett, C. A., F. C. Bellrose, H. H. Burgess, A. S. Hawkins, and J. A. Kadlec. 1986. Declining productivity of trumpeter swans at Red Rock Lakes National Wildlife Refuge, Lima, Montana. *Trumpeter Swan Society Conference* 9: 124–131.

Innes, D. 1994. Trumpeter swan Pacific Coast Population status in the Comox area of the Vancouver Island, British Columbia. *Trumpeter Swan Society Conference* 14: 72–73.

Intini, K. 2019. Ontario trumpeter swan program status update. Paper presented at 25th Trumpeter Swan Society Conference, November 19–20, 2019, Alton, Illinois.

Intini, K., J. Kee, B. Kingdon, and R. W. Kingdon. 2016. Ontario trumpeter swan (*Cygnus buccinator*) status update and results of the 2015 North American trumpeter swan survey (abstract). *Trumpeter Swan Society Conference* 24 (unpaged).

Ivey, G. L. 1990. Population status of trumpeter swans in southeastern Oregon. *Trumpeter Swan Society Conference* 11: 118–122.

———. 2016. Status of trumpeter swans in Oregon (abstract). *Trumpeter Swan Society Conference* 24 (unpaged).

———. 2019. The Oregon Project and evidence of increasing connectivity in the western US flocks. Paper presented at 25th Trumpeter Swan Society Conference, November 19–20, 2019, Alton, Illinois.

Ivey, G. L., and C. G. Carey. 1991. A plan to enhance Oregon's trumpeter swan population. *Trumpeter Swan Society Conference* 12: 18–23.

Ivey, G. L., and J. E. Cornely. 2007. Survival analysis of Malheur National Wildlife Refuge trumpeter swans (abstract). *Trumpeter Swan Society Conference* 20: 106.

Ivey, G. L., and J. E. Cornely 2016. Conservation needs of trumpeter swans in the western United States (abstract). *Trumpeter Swan Society Conference* 24 (unpaged).

Ivey, G. L., M. J. St. Louis, and B. D. Bales. 2001. The status of the Oregon trumpeter swan program. *Trumpeter Swan Society Conference* 17: 108–113.

Jobes, C. R. 1990. Growth characteristics of trumpeter swan cygnets from different populations. *Trumpeter Swan Society Conference* 11: 164.

Johnsgard, P. A. 1978. The triumphant trumpeter. *Natural History* 87(9): 72–77. https://digitalcommons.unl.edu/biosciornithology/18/

———. 1982. *Teton Wildlife: Observations by a Naturalist.* Boulder: Colorado Associated University Press. 128 pp. https://digitalcommons.unl.edu/biosciornithology/52/

———. 2012. *Squaw Creek National Wildlife Refuge: Gem of the Missouri Valley.* Prairie Fire, November 2012, pp. 12–13. http://www.prairiefirenewspaper.com/2012/11/squaw-creek-national-wildlife-refuge-gem-of-the-missouri-valley

———. 2013. *Yellowstone Wildlife: Ecology and Natural History of the Greater Yellowstone Ecosystem.* Boulder: University Press of Colorado. 239 pp.

———. 2018. *Wyoming Wildlife: A Natural History.* Lincoln: University of Nebraska-Lincoln DigitalCommons and Zea Books. 244 pp. https://digitalcommons.unl.edu/zeabook/73/

Johnson, W. C. 1991. Michigan's trumpeter swan restoration program. *Trumpeter Swan Society Conference* 123: 108–110.

———. 1999a. Michigan 1996 trumpeter swan update. *Trumpeter Swan Society Conference* 16: 18–20.

———. 1999b. Observations of territorial conflict between trumpeter swans and mute swans in Michigan. *Trumpeter Swan Society Conference* 16: 134–136.

———. 2007. Status of the Michigan population of trumpeter swans, 2005. *Trumpeter Swan Society Conference* 20: 20–21.

Jordan, M. 1986. A summary of the distribution and status of trumpeter swans in Washington State. *Trumpeter Swan Society Conference* 9: 67–70.

———. 1990. A summary of the status of trumpeter swans in Washington State. *Trumpeter Swan Society Conference* 11: 113–116.

———. 1991. Trumpeter and tundra swan survey in western Washington and Oregon—January 1989. *Trumpeter Swan Society Conference* 12: 14–17.

Jordan, M., L. N. Gillette, R. E. Shea. 2000. Summary of trumpeter swan priorities identified during the 17th Trumpeter Swan Society Conference September 1999. *Trumpeter Swan Society Conference* 17: 179–180.

Kaye, R., and L. Shandruk. 1992. Elk Island National Park trumpeter swan reintroduction—1990. *Trumpeter Swan Society Conference* 13: 22–30.

Kearns, L. 2016. The recovery of trumpeter swans in Ohio (abstract). *Trumpeter Swan Society Conference* 24 (unpaged).

Kearns, L. 2019. Trumpeter swan status and management actions in Ohio. Paper presented at 25th Trumpeter Swan Society Conference, November 19-20, 2019, Alton, Illinois.

Killaby, M. 1988. Trumpeter swan habitation and proposed management in Saskatchewan. *Trumpeter Swan Society Conference* 10: 47–48.

Kilpatrick, D. 2007. Translocating trumpeter swans from the Rocky Mountain Population: Habitat, movement, and survival. MS thesis. Moscow: University of Idaho.

Kilpatrick, D., K. P. Reese, L. Hanauska-Brown, and T. Hemker. 2007. Trumpeter swan translocation project 2001-2005 in Idaho: Survival and movement (abstract). *Trumpeter Swan Society Conference* 20: 98–99.

King, J. G. 1986. Managing to have wild trumpeter swans on a continent exploding with people. *Trumpeter Swan Society Conference* 9: 119–123.

———. 1988. New goals for the second half century of trumpeter swan restoration. *Trumpeter Swan Society Conference* 10: 118–120.

———. 1994. Pacific Coast trumpeter swans in the 21st century. *Trumpeter Swan Society Conference* 14: 3–4.

———. 1996. Trying to understand what swans think about, especially winter habitat. *Trumpeter Swan Society Conference* 15: 41–43.

———. 2000. Are Alaska's wild swans safe? *Trumpeter Swan Society Conference* 17: 4–5.

———. 2007. Comparison of 290 photos of wild swan nests. *Trumpeter Swan Society Conference* 20: 130–135.

King, J. G., and B. Conant. 1981. The 1980 census of trumpeter swans on Alaskan nesting habitats. *American Birds* 35(5): 789–793.

King, J. G., R. Ritchie, B. Cooper, and H. McMahan. 1992. Flying with the swans through Alaska's great mountains. *Trumpeter Swan Society Conference* 13: 165–168.

King, R. J. 1985. *Trumpeter Swan* (Cygnus buccinator) *Movements from the Tanana Valley, Alaska.* Fairbanks, AK: US Fish and Wildlife Service.

———. 1988. Progress report on impact of collecting trumpeter swan (*Cygnus buccinator*) eggs in Minot Flats, Alaska—1986. *Trumpeter Swan Society Conference* 10: 89–95.

———. 1990. Impacts on trumpeter swans (*Cygnus buccinator*) from egg collection activities in the Minto Flats, Alaska, 1987 (abstract). *Trumpeter Swan Society Conference* 11: 106.

———. 1991a. Migration and wintering resightings of trumpeter swans from central Alaska (abstract). *Trumpeter Swan Society Conference* 12: 10.

———. 1991b. Impacts on trumpeter swans from egg collection activities in Minto Flats, Alaska. *Trumpeter Swan Society Conference* 12: 28–44.

———. 1992. Management problems on the breeding grounds and strategies to resolve them. *Trumpeter Swan Society Conference* 13: 160–164.

———. 1994. Trumpeter swan movements from Minto Flats, Alaska: 1982–92. *Trumpeter Swan Society Conference* 14: 19–36.

Kittelson, S. M. 1990a. An update of the Minnesota Department of Natural Resources trumpeter swan restoration project. *Trumpeter Swan Society Conference* 11: 50–52.

———. 1990b. Avicultural techniques used in Minnesota trumpeter swan restoration. *Trumpeter Swan Society Conference* 11: 174–176.

———. 1991. An update of Minnesota's trumpeter swan restoration efforts—the captive rearing program. *Trumpeter Swan Society Conference* 12: 95–96.

Kittelson, S. M., and P. Hines. 1992. Minnesota Department of Natural Resources trumpeter swan restoration project status report. *Trumpeter Swan Society Conference* 13: 109–113.

Kraege, D. 1996. Pacific Coast Population—status, trends, and management issues. *Trumpeter Swan Society Conference* 15: 3–8.

Kraft, R. H. 1990. Status report of the Lacreek National Wildlife Refuge, South Dakota, trumpeter swan flock management plan. *Trumpeter Swan Society Conference* 11: 58–61.

———. 1991a. Status report of the Lacreek trumpeter swan flock. *Trumpeter Swan Society Conference* 12: 88–90.

———. 1991b. A proposal to restore winter migration in trumpeter swans by establishing breeding pairs in the wintering area. *Trumpeter Swan Society Conference* 12: 142.

———. 1996. Observations of trumpeter swan behavior and management techniques. *Trumpeter Swan Society Conference* 15: 91–98.

———. 2001. Status report of the High Plains flock for 2000. *North American Swans* 30(1): 32–37.

———. 2004. Status report of the Lacreek trumpeter swan flock for 2002. *Trumpeter Swan Society Conference* 19: 148–154.

Lamb, G. 2004. Swan habitat conservation and partnerships on Long Beach Peninsula, Washington State (abstract). *Trumpeter Swan Society Conference* 19: 9.

LaMontagne, J. M., L. J. Jackson, and R. M. R. Barclay. 2003a. Characteristics of ponds used by trumpeter swans in a spring migration stopover area. *Canadian Journal of Zoology* 81: 1791–1798.

LaMontagne, J. M., L. J. Jackson, and R. M. R. Barclay. 2003b. Compensatory growth responses of *Potamogeton pectinatus* to foraging by migrating trumpeter swans in spring stopover areas. *Aquatic Botany* 76: 235–244.

LaMontagne, J. M., L. J. Jackson, and R. M. R. Barclay. 2004. Energy balance of trumpeter swans at stopover areas during spring migration. *Northwestern Naturalist* 85: 104–110.

LaMontagne, J. M., R. M. R. Barclay, and L. J. Jackson. 2001. Trumpeter swan behaviour at spring-migration stopover areas in southern Alberta. *Canadian Journal of Zoology* 79: 2036–2042.

Lance, E. W., and E. J. Mallek. 2004. One year of satellite telemetry data for four Alaskan trumpeter swans. *Trumpeter Swan Society Conference* 19: 29–38.

Lawrence, S. 1999. The Monticello swans. *Trumpeter Swan Society Conference* 16: 32–38.

———. 2007. The trumpeter swans of Monticello, Minnesota. *Trumpeter Swan Society Conference* 20: 153–155.

Lemaster, D. B. 1981. Foraging ecology of a population of trumpeter swans wintering in southeast Alaska. MS thesis. Pocatello: Idaho State University.

Linck, M. H. 1999. Advantages and disadvantages of a wintering congregation of trumpeter swans on the Mississippi River, Monticello (Wright County), Minnesota. *Trumpeter Swan Society Conference* 16: 30–31.

Linck, M. H., K. Rowe, and J. Mosby. 2007. The trumpeter swans of Heber Springs, Cleburne County, Arkansas. *Trumpeter Swan Society Conference* 20: 42–44.

Lockman, D. C. 1986. North American Management Plan as it pertains to the Rocky Mountain Population. *Trumpeter Swan Society Conference* 9: 4.

——. 1988. Wyoming trumpeter swan progress report. *Trumpeter Swan Society Conference* 10: 60–63.

——. 1990a. Wyoming trumpeter swan program status. *Trumpeter Swan Society Conference* 11: 9–11.

——. 1990b. Trumpeter swan mortality in Wyoming, 1982–1987. *Trumpeter Swan Society Conference* 11: 12–13.

——. 1990c. Rocky Mountain Trumpeter Swan Population Subcommittee report (abstract). *Trumpeter Swan Society Conference* 11: 38.

——. 1990d. Rocky Mountain Trumpeter Swan Population range expansion project, 1988–1993. *Trumpeter Swan Society Conference* 11: 40–43.

——. 1991. Strategies tested in Wyoming for trumpeter swan range expansion—1989 progress report. *Trumpeter Swan Society Conference* 12: 196–199.

Lockman, D. C., C. D. Mitchell, B. Reiswig, and R. S. Gale. 1990. Identifying potential winter habitat for trumpeter swans. *Trumpeter Swan Society Conference* 11: 20–22.

Lockman, D. C., R. Wood, B. Smith, B. Raynes, and D. Childress. 1986. Progress report: Rocky Mountain Trumpeter Swan Population—Wyoming flock, 16 September 1983 through 15 September 1984. *Trumpeter Swan Society Conference* 9: 29–47.

Lockman, D. C., R. Wood, H. H. Burgess, R. Burgess, and H. Smith. 1990. Trumpeter swan seasonal habitat use in western Wyoming. *Trumpeter Swan Society Conference* 11: 14–19.

Long, W. M., and D. Stevenson. 1999. Trumpeter swan range restoration in Wyoming. *Trumpeter Swan Society Conference* 16: 66–71.

Looft, J. A. 2014. Estimating trumpeter swan (*Cygnus buccinator*) populations in Alberta and response to disturbance. MS thesis. Edmonton: University of Alberta.

Luebbert, J. 2000. The science of migration and navigation: Considerations for trumpeter swan (*Cygnus buccinator*) translocations. *Trumpeter Swan Society Conference* 17: 164–169.

Lueck, C. 1991. The agonistic behavior of nesting trumpeter swans to other waterfowl. *Trumpeter Swan Society Conference* 12: 163–169.

Lumbis, K. 1988. Wetland habitat management in the Grande Prairie region of northwestern Alberta. *Trumpeter Swan Society Conference* 10: 22–27.

Lumsden, H. G. 1988a. Productivity of trumpeter swans in relation to condition. *Trumpeter Swan Society Conference* 10: 150–154.

——. 1988b. The food of trumpeter swan cygnets in Ontario. *Trumpeter Swan Society Conference* 10: 155–157.

——. 1992. Trumpeter swans once bred on the Atlantic Coast. *Trumpeter Swan Society Newsletter* 21(2): 8.

——. 2000. Induced migration—its origins and history. *Trumpeter Swan Society Conference* 17: 158–162.

——. 2001. A survey of trumpeter swans in the Kenora District of Ontario. *North American Swans* 30(1): 19–20.

——. 2002a. Laying and incubation behavior of captive trumpeter swans. Special Publication 1: Proceedings of the Fourth International Swan Symposium 2001 (E. C. Rees, S. L. Earnst, and J. Coulson, eds.). *Waterbirds* 25: 293–295.

——. 2002b. Hatchability of eggs from captive trumpeter swans. *North American Swans* 31: 2–4.

——. 2002c. The trumpeter swan restoration program in Ontario—2001. *North American Swans* 31: 16–18.

——. 2004. The trumpeter swan restoration program in Ontario—2002. *Trumpeter Swan Society Conference* 19: 155–158.

———. 2007a. The inventory of trumpeter swans in Ontario in 2005. *Trumpeter Swan Society Conference* 20: 11.

———. 2007b. Migration of Ontario trumpeter swans. *Trumpeter Swan Society Conference* 20: 37-40.

———. 2012. Recent history of trumpeter swans in Ontario and Quebec and their status in 2010-2011. *Ontario Birds* 30: 109-119.

———. 2013. Isotherms and winter distribution of trumpeter swans. *Ontario Birds* 31: 46-50.

———. 2016a. Colour morphs, downy and juvenile plumages of trumpeter and mute swans. *Ontario Birds* 34(2): 198-204.

———. 2016b. Trumpeter swans and mute swans compete for space in Ontario. *Ontario Birds* 34: 14-23.

———. 2017. Egg carrying by a trumpeter swan. *Ontario Birds* 35(1): 17-19.

Lumsden, H. G., and M. C. Drever. 2002. Overview of the trumpeter swan reintroduction program in Ontario, 1982-2000. Special Publication 1: Proceedings of the Fourth International Swan Symposium 2001 (E. C. Rees, S. L. Earnst, and J. Coulson, eds.). *Waterbirds* 25: 301-312.

Lumsden, H. G., D. Compton, J. Johnson, S. Kittelson, P. Hines, S. Matteson, and J. Smith. 1994. Trumpeter swan restoration in the Midwest. *Trumpeter Swan Society Conference* 14: 145-149.

Lumsden, H. G., D. McLachlin, and P. Nash. 1988. Restoration of trumpeter swans in Ontario. *Trumpeter Swan Society Conference* 10: 110-116.

Luszcz, D. 2000. Status of Atlantic Flyway Trumpeter Swan Management Plan. *Trumpeter Swan Society Conference* 17: 9-10.

Mackay, J. 2004. The Ruby Lake trumpeter swan flock: Its history, current status, and future. *Trumpeter Swan Society Conference* 19: 121-127.

Mackay, R. H. 1957. Movements of trumpeter swans shown by band returns and observations. *Condor* 59: 339.

———. 1988. Trumpeter swan investigations, Grand Prairie, Alberta, 1953-75. *Trumpeter Swan Society Conference* 10: 5-10.

Maj, M. 1983. Analysis of trumpeter swan habitat on the Targhee National Forest of Idaho and Wyoming. MS thesis. Bozeman: Montana State University.

———. 1986. Targhee National Forest trumpeter swans. *Trumpeter Swan Society Conference* 9: 52-53.

Marsolais, J. V. 1994. The genetic status of trumpeter swan (*Cygnus buccinator*) populations (abstract). *Trumpeter Swan Society Conference* 14: 162-164.

Marsolais, J. V., and B. N. White. 1997. Genetic considerations for the reintroduction of trumpeter swans in Ontario. *North American Swans* 26: 18-22.

Matteson, S. W. 1991. Wisconsin's trumpeter swan recovery program. *Trumpeter Swan Society Conference* 12: 106-107.

———. 2019. Wisconsin's trumpeter swan recovery program: A 30-year retrospective (1989-2019) on research, management, and collaboration. Paper presented at 25th Trumpeter Swan Society Conference, November 19-20, 2019, Alton, Illinois.

Matteson, S. W., S. Craven, and D. Compton. 1995. *The Trumpeter Swan*. Madison: University of Wisconsin Extension Publication G3647. 8 pp.

Matteson, S. W., P. F. Manthey, M. J. Mossman, and L. M. Hartman. 2007. Wisconsin trumpeter swan recovery program: Progress toward restoration, 1987-2005. *Trumpeter Swan Society Conference* 20: 11-18.

Matteson, S. W., P. Manthey, M. J. Mossman, L. Hartman, and E. Diebold. 2014. Wisconsin's trumpeter swan recovery program: A 25-year retrospective (1987-2012) on techniques, habitat characteristics, management, threats, and collaboration (abstract). *Trumpeter Swan Society Conference* 23 (unpaged).

Matteson, S. W., M. J. Mossman, and L. M. Hartman. 1996. Wisconsin's trumpeter swan restoration efforts, 1987-1994. *Trumpeter Swan Society Conference* 15: 73-85.

Matteson, S. W., M. J. Mossman, R. Jurewicz, and E. Diebold. 1992. Collection, transport, and hatching success of Alaskan trumpeter swan eggs, 1989–90, and status of Wisconsin's trumpeter swan recovery program. *Trumpeter Swan Society Conference* 13: 105-108.

McCormick, K. J. 1986. Status of trumpeter swans in the Northwest Territories. *Trumpeter Swan Society Conference* 9: 22-27.

McEneaney, T. 1986. Effects of water flow fluctuations, icing, and recreationists on the distribution of wintering trumpeter swans in the Tristate region. *Trumpeter Swan Society Conference* 9: 53-54.

———. 1991. Status of the trumpeter swan in Yellowstone National Park. *Trumpeter Swan Society Conference* 12: 193-194.

———. 1996. Trumpeter swan management within and beyond park boundaries. *Trumpeter Swan Society Conference* 15: 151-155.

———. 2005. Rare color variants of the trumpeter swan. *Birding* 37: 148-154.

McKelvey, R. W. 1981. Some aspects of the winter foraging ecology of trumpeter swans at Port Alberni and Comox Harbour, British Columbia. MS thesis. Burnaby, BC: Simon Fraser University.

———. 1986a. An overview of the North American Management Plan for Trumpeter Swans as it pertains to the Pacific Coast population. *Trumpeter Swan Society Conference* 9: 62-64.

———. 1986b. Notes on the status of trumpeter swans in British Columbia and the Yukon Territory, and on grazing studies at Comox Harbour, British Columbia. *Trumpeter Swan Society Conference* 9: 65-67.

———. 1986c. Guidelines for trumpeter swan restoration in Canada. *Trumpeter Swan Society Conference* 9: 111-112.

———. 1988. The 1985 survey of trumpeter swans in British Columbia and Yukon. *Trumpeter Swan Society Conference* 10: 145.

———. 1991. The status of trumpeter swans wintering in southwestern British Columbia in 1989. *Trumpeter Swan Society Conference* 12: 12-13.

———. 1992. Canadian involvement in management of trumpeter swans. *Trumpeter Swan Society Conference* 13: 173-175.

Mckelvey, R. W., and A. C. MacNeill. 1981. Mortality factors of wild swans in Canada. Pp. 312-318 in *Proceedings of the Second International Swan Symposium* (G. V. T. Matthews and M. Smart, eds.). Slimbridge, UK: International Waterfowl and Wetland Research Bureau.

Mckelvey, R. W., and C. Burton. 1983. A possible migration route for trumpeter swans (*Cygnus buccinator*) in British Columbia. *Canadian Wildlife Service Progress Report* 138.

McKelvey, R. W., and N. A. M. Verbeek. 1988. Habitat use, behavior, and management of trumpeter swans, *Cygnus buccinator*, wintering at Comox, British Columbia. *Canadian Field-Naturalist* 102: 434-441.

Meng, A., and D. T. Parkin. 1993. Genetic differentiation in natural populations of swans revealed by DNA fingerprinting. *Acta Zoologica Sinica* 39(2): 209-216.

Miller, P. 2109. Incubation ecology of trumpeter swans (*Cygnus buccinator*). Paper presented at 25th Trumpeter Swan Society Conference, November 19-20, 2019, Alton, Illinois.

Mitchell, C. D. 1990. Efficiency of techniques for feeding wintering trumpeter swans. *Trumpeter Swan Society Conference* 11: 170-173.

———. 1991. Update on trumpeter swans in Montana—1988-89. *Trumpeter Swan Society Conference* 12: 189-192.

———. 1994. Trumpeter swan research needs. *Trumpeter Swan Society Conference* 14: 155-161.

———. 2004. Trumpeter swan restoration at Grays Lake National Wildlife Refuge, Idaho. *Trumpeter Swan Society Conference* 19: 108-115.

Mitchell, C. D., and J. J. Rotella. 1997. Brood amalgamation in trumpeter swans. *Wildfowl* 48: 1-5.

Mitchell, C. D., and L. Shandruk. 1992. Rocky Mountain Population of trumpeter swans: Status, distribution, and movements. *Trumpeter Swan Society Conference* 13: 3-18.

Mitchell, C. D., and M. W. Eichholz. 2019. Trumpeter swan (*Cygnus buccinator*). Version 3.0. *Birds of North America* (A. F. Poole, ed.). Ithaca, NY: Cornell Laboratory of Ornithology. https://birdsna.org/Species-Account/bna/species/truswa/introduction

Mitchell, C. D., R. Shea, D. C. Lockman, and J. R. Balcomb. 1991. Demographic analyses of a trumpeter swan *Cygnus cygnus buccinator* population in western USA. P. 142 in *Wildfowl: Supplement Number 1, Third IWRB International Swan Symposium*, Oxford, England, December 9-13, 1989 (J. Sears and P. J. Bacon, eds.). Slimbridge, UK: International Waterfowl and Wetlands Research Bureau.

Monnie, J. B. 1966. Reintroduction of the trumpeter swan to its former breeding range. *Journal of Wildlife Management* 30: 691-696.

Monson, M. A. 1956. Nesting of trumpeter swans in the lower Copper River basin, Alaska. *Condor* 58: 444-445.

Moriarty, J. J. 1991. Feather stress in trumpeter swans. *Trumpeter Swan Society Conference* 12: 152.

Morrison, K. F. 1990. Number and age composition of trumpeter swans wintering on the east coast of Vancouver Island, British Columbia, 1983-1988. *Trumpeter Swan Society Conference* 11: 107-112.

Moser, T. J. 2006. The 2005 North American Trumpeter Swan Survey. Unpublished report. Denver, CO: US Fish and Wildlife Service, Division of Migratory Bird Management.

Moser, T. J., J. E. Cornely, and F. Caithamer. 2008. The North American Quintennial Trumpeter Swan Survey, 1968-2005 (abstract). *Trumpeter Swan Society Conference* 21 (unpaged).

Mossman, M. J., and S. W. Matteson. 1990. Trumpeter swan status report for Wisconsin. *Trumpeter Swan Society Conference* 11: 53-55.

Munoz, R. 2000. Role of Southeast Idaho National Wildlife Refuge Complex in the Rocky Mountain Population trumpeter swan project. *Trumpeter Swan Society Conference* 17: 107-108.

Murase, M. 1993. The first record of the trumpeter swan *Cygnus buccinators* for Japan. *Japanese Journal of Ornithology* 41(2): 51-55.

Myhr, R. 1996. Preserving trumpeter swan habitat in the San Juan Islands: Perhaps an example for other land trusts. *Trumpeter Swan Society Conference* 15: 136-138.

Nelson, H. K. 1994. The Trumpeter Swan Society's position on swan hunting. *Trumpeter Swan Society Conference* 14: 137-142.

———. 1999. Development of a management plan for the Interior Population of trumpeter swans. *Trumpeter Swan Society Conference* 16: 27-29.

Neptune, B. 2007. The nesting trumpeter swans of Dawn, Missouri. *Trumpeter Swan Society Conference* 20: 158-159.

Niethammer, K. R., D. Gomez, and S. M. Linneman. 1994. Termination of winter feeding of trumpeter swans at Red Rock Lakes National Wildlife Refuge—a progress report. *Trumpeter Swan Society Conference* 14: 118-121.

North, M. R. 2007. The earliest historical records of trumpeter swans—extralimital to today's distribution. *Trumpeter Swan Society Conference* 20: 148-150.

Norton, M., and G. Beyersbergen. 2001. 2000 survey of trumpeter swans in Alberta, Saskatchewan, Manitoba, and the Northwest Territories. *North American Swans* 30(1): 25-31.

Olson, D. 2001. 2001 Midwinter survey: Rocky Mountain Population of trumpeter swans. *North American Swans* 30(1): 38-42.

———. 2002. 2002 Midwinter survey: Rocky Mountain Population of trumpeter swans. *North American Swans* 31(1): 12-15.

———. 2004. Analysis of winter satellite telemetry locations from trumpeter swans marked and released at Red Rock Lakes National Wildlife Refuge, Montana (abstract). *Trumpeter Swan Society Conference* 19: 95.

———. 2011. Winter and summer use of the core and expansion areas in the Greater Yellowstone ecosystem by swans of the Rocky Mountain Population (abstract). *Trumpeter Swan Society Conference* 22 (unpaged).

———. 2016. *Trumpeter Swan Survey of the Rocky Mountain Population, US Breeding Segment, Fall 2015*. Lakewood, CO: US Fish and Wildlife Service, Mountain Prairie Region. 39 pp.

———. 2019a. Results of the 2019 trumpeter swan survey of the RMP, US Breeding Segment. Paper presented at 25th Trumpeter Swan Society Conference, November 19-20, 2019, Alton, Illinois.

———. 2019b. Status of the High Plains flock of trumpeter swans. Paper presented at 25th Trumpeter Swan Society Conference, November 19-20, 2019, Alton, Illinois.

Olson, D., and R. Gregory. 2007. Status of trumpeter swans (*Cygnus buccinator*) at Seney National Wildlife Refuge (1991-2005) (abstract). *Trumpeter Swan Society Conference* 20: 22.

Olson D., B. Long, and C. D. Mitchell. 2015. Geographic variation in trumpeter swan *Cygnus buccinator* clutch size and egg weights. *Wildfowl* 65: 133-142.

Olson D., J. Warren, and T. Reed. 2009. Satellite-tracking the seasonal locations of trumpeter swans *Cygnus buccinator* from Red Rock Lakes National Wildlife Refuge, Montana, USA. *Wildfowl* 59: 3-16.

Olson, D., R. G. Corace III, D. L. McCormick, and V. Cavelieri. 2007. Status of trumpeter swans (*Cygnus buccinator*) at Seney National Wildlife Refuge (1991-2005). *North American Swans* 33: 22-23.

Orme, M. L., and R. E. Shea. 2000. Trumpeter swan nesting habitat on the Targhee National Forest. *Trumpeter Swan Society Conference* 17: 95-102.

Oyler-McCance, S. J., and T. W. Quinn. 2004. Comparison of trumpeter swan populations using nuclear and mitochondrial genetic markers (abstract). *Trumpeter Swan Society Conference* 19: 119.

Oyler-McCance, S. J., F. A. Ransler, L. K. Berkman, and T. W. Quinn. 2007. A rangewide population genetic study of trumpeter swans. *Conservation Genetics* 8: 1339-1353.

Pacific Flyway Council. 2002. *Pacific Flyway Implementation Plan for the Rocky Mountain Population of Trumpeter Swans*. Portland, OR: Pacific Flyway Study Committee, US Fish and Wildlife Service.

Page, R. D. 1976. The ecology of the trumpeter swan on Red Rock Lakes National Wildlife Refuge, Montana. PhD diss. Missoula: University of Montana.

Patla, S., and R. Oakleaf. 2004. Summary and update of trumpeter swan range expansion efforts in Wyoming, 1988-2003. *Trumpeter Swan Society Conference* 19: 116-118.

Patton, K., E. Butterworth, D. Falk, A. Leach, and C. Smith. 2004. Records of trumpeter swans in the Ducks Unlimited Canada western boreal program. *Trumpeter Swan Society Conference* 19: 44-49.

Peiplow, N. 2019. *Peterson Field Guide to Bird Sounds of Western North America*. Boston: Houghton Mifflin.

Pelizza, C. A. 2000. Winter site selection characteristics, genetic composition, and mortality factors of the High Plains flock of trumpeter swans. *Trumpeter Swan Society Conference* 17: 29-34.

———. 2001. Winter ecology of the trumpeter swan *Cygnus buccinator* in the northern Great Plains. MS thesis. Vermillion: University of South Dakota.

Pelizza, C. A., and H. B. Britten. 2002. Isozyme analysis reveals genetic differences between three trumpeter swan populations. Special Publication 1: Proceedings of the Fourth International Swan Symposium 2001 (E. C. Rees, S. L. Earnst, and J. Coulson, eds.). *Waterbirds* 25: 355-359.

Pichner, J. 1991. Trumpeter swan multiple and continuous clutching—a summary. *Trumpeter Swan Society Conference* 12: 148-151.

Pichner, J., N. Reindl, and B. Geiszler. 1990. Double clutching of trumpeter swans (*Cygnus cygnus buccinator*) at the Minnesota Zoological Garden. *Trumpeter Swan Society Conference* 11: 177-178.

Pichner, J., S. Kittelson, and P. Hines. 1992. Survival of hand-reared and parent-reared trumpeter swans (*Cygnus buccinator*) in the Minnesota Department of Natural Resources restoration project. *Trumpeter Swan Society Conference* 13: 114-118.

Price, A. L., J. Kyler, and R. L. Studebaker. 1999. Public participation in the restoration of the trumpeter swans within the Interior Population. *Trumpeter Swan Society Conference* 16: 39-43.

Proffitt, K. M., T. McEneaney, P. J. White, and R. A. Garrott. 2009. Trumpeter swan abundance and growth rates in Yellowstone National Park. *Journal of Wildlife Management* 73: 728-736.

———. 2010. Productivity and fledging success of trumpeter swans in Yellowstone National Park, 1987-2007. *Waterbirds* 33: 341-348.

Quirck, W. A., III. 2014. Density and productivity comparisons between two distinctive breeding flocks of trumpeter swans in southcentral Alaska (abstract). *Trumpeter Swan Society Conference* 23 (unpaged).

Ransler, F. A., T. W. Quinn, and S. J. Oyler-McCance. 2011. Genetic consequences of trumpeter swan (*Cygnus buccinator*) reintroductions. *Conservation Genetics* 12(1): 257-268.

Rasmussen, P. J. 2007. Managing Monticello trumpeter swans and power line issues (abstract). *Trumpeter Swan Society Conference* 20: 34-36.

Rees, J. 1981. *Historical and Demographic Analysis of a Trumpeter Swan Introduction on Turnbull National Wildlife Refuge.* Cheney, WA: US Fish and Wildlife Service.

Reiswig, B. 1986. Status of the Tristate subpopulation and the Rocky Mountain winter population of trumpeter swans. *Trumpeter Swan Society Conference* 9: 7-10.

———. 1988a. The Trumpeter Swan Society's Red Rock Lakes Study Committee recommendations: A US Fish and Wildlife Service update. *Trumpeter Swan Society Conference* 10: 50-53.

———. 1988b. A review of wintering Rocky Mountain trumpeter swan population survey estimates: 1977-1986. *Trumpeter Swan Society Conference* 10: 57-59.

———. 1990. Trumpeter swan management—Montana overview. *Trumpeter Swan Society Conference* 11: 2-3.

Reiswig, B., and C. D. Mitchell. 1996. Rocky Mountain Population of trumpeter swans: Status, trends, problems, outlook. *Trumpeter Swan Society Conference* 15: 9-17.

Ripley, L. 1984. *Alberta/BC Cooperative Trumpeter Swan Restoration Program, 1983.* Brooks: Alberta Department of Forestry, Lands and Wildlife.

———. 1985. *Trumpeter Swan Restoration Program, 1984.* Brooks: Alberta Department of Forestry, Lands and Wildlife.

Rocky Mountain Population Trumpeter Swan Subcommittee. 1990. Contingency plan for management of wintering trumpeter swans in the vicinity of Harriman State Park, Idaho. *Trumpeter Swan Society Conference* 11: 44-45.

———. 2017. *Management Plan for the Rocky Mountain Population of Trumpeter Swans.* Pacific Flyway Council, c/o US Fish and Wildlife Service, Division of Migratory Bird Management, Vancouver, Washington, DC. 50 pp.

Rogers, P. M., and D. A. Hammer. 1978. Ancestral Breeding and Wintering Ranges of the Trumpeter Swan (*Cygnus buccinator*) in the Eastern United States. Unpublished report. Knoxville: Tennessee Valley Authority.

Rolston, G. 1994. Land use conflicts in the Comox Valley. *Trumpeter Swan Society Conference* 14: 77-78.

Schmidt, J. H., M. S. Lindberg, D. S. Johnson, B. Conant, and J. G. King. 2008a. Growth of trumpeter swan population in Alaska from 1968-2006 and the effects of climate change on habitat occupancy (abstract). *Trumpeter Swan Society Conference* 21 (unpaged).

———. 2008b. Evidence of trumpeter swan population growth using Bayesian hierarchical models. *Journal of Wildlife Management* 73: 720-727.

Schmidt, J. H., M. S. Lindberg, D. S. Johnson, and D. L. Verbyla. 2011. Season length influences breeding range dynamics of trumpeter swans *Cygnus buccinator*. *Wildlife Biology* 17: 364-372.

Schmidt, P. 2000. Challenges in conserving swans and other migratory birds into the next millennium. *Trumpeter Swan Society Conference* 17: 41-43.

Schorger, A. W. 1964. The trumpeter swan (*Olor buccinator*) as a breeding bird in Minnesota, Wisconsin, Illinois, and Indiana. *Wilson Bulletin* 76(4): 331-338.

Shandruk, L. J. 1986. Draft proposal: A long range habitat management strategy for the Interior Canada subpopulation of trumpeter swans. *Trumpeter Swan Society Conference* 9: 14-22.

———. 1988a. Status of trumpeter swans in the southern Mackenzie District, Northwest Territories. *Trumpeter Swan Society Conference* 10: 42-46.

———. 1988b. Elk Island National Park trumpeter swan transplant pilot project—final report. *Trumpeter Swan Society Conference* 10: 66-77.

————. 1988c. A survey of trumpeter swan breeding habitats in Alberta, Saskatchewan, and northeastern British Columbia. *Trumpeter Swan Society Conference* 10: 131–144.

Shandruk, L. J., and G. Holton. 1986. Status report: A pilot project to transplant trumpeter swans into Elk Island National Park, Alberta. *Trumpeter Swan Society Conference* 9: 11–13.

Shandruk, L. J., and K. J. McCormick. 1990. Status of trumpeter swans in the southern Mackenzie District, Northwest Territories, 1986 and 1987. *Trumpeter Swan Society Conference* 11: 23–27.

————. 1991. Status of the Grande Prairie and Nahanni trumpeter swan flocks. *Trumpeter Swan Society Conference* 12: 181–183.

Shandruk, L. J., and R. Kaye. 1991. Elk Island National Park trumpeter swan reintroduction—1989 progress report. *Trumpeter Swan Society Conference* 12: 184–188.

Shandruk, L. J., and T. Winkler. 1990. Elk Island National Park trumpeter swan reintroduction, 1987 progress report. *Trumpeter Swan Society Conference* 11: 28–32.

Sharp, D. E., and J. B. Bortner. 1991. North American Waterfowl Management Plan. *Trumpeter Swan Society Conference* 12: 122–126.

Shea, R. E. 1979. The ecology of trumpeter swans in Yellowstone National Park and vicinity. MS thesis. Missoula: University of Montana.

————. 1992. Response of trumpeter swans to trapping at Red Rock Lakes National Wildlife Refuge, Montana, and Harriman State Park, Idaho, winter 1991. *Trumpeter Swan Society Conference* 13: 31–37.

————. 1999. Recent changes in distribution and abundance of the Rocky Mountain Population of trumpeter swans. *Trumpeter Swan Society Conference* 16: 49–55.

————. 2000. Rocky Mountain trumpeter swans: Current vulnerability and restoration potential. *Trumpeter Swan Society Conference* 17: 74–81.

————. 2004. Status of trumpeter swans nesting in the western United States and management issues. *Trumpeter Swan Society Conference* 19: 85–94.

Shea, R. E., E. O. Garton, and I. J. Ball. 2013. *The History, Ecology, and Management of the Rocky Mountain Population of Trumpeter Swans (1931–86). North American Swans, Bulletin of the Trumpeter Swan Society* 34(1): 1–278. 2nd ed. https://www.bookdepository.com/History-Ecology-Management-Rocky-Mountain-Population-Trumpeter-Swans-1931-86-Ruth-E-Shea/9781484867211

Shea, R. E., H. K. Nelson, L. N. Gillette, J. G. King, and D. K. Weaver. 2002. Restoration of trumpeter swans in North America: A century of progress and challenges. Special Publication 1: Proceedings of the Fourth International Swan Symposium 2001. (E. C. Rees, S. L. Earnst, and J. Coulson, eds.) *Waterbirds* 25: 296–300.

Shea, R. E., R. C. Drewien, and C. S. Peck. 1994. Overview of efforts to expand the range of the Rocky Mountain Population of trumpeter swans. *Trumpeter Swan Society Conference* 14: 111–117.

Sheehan, B. 1988. Early history of trumpeter swans in the Grande Prairie area. *Trumpeter Swan Society Conference* 10: 2–4.

Shepherd, P. E. K. 1962. An ecological reconnaissance of the trumpeter swan in south central Alaska. MS thesis. Pullman: Washington State University.

Shields, E. 2019. Loss of an icon: Can trumpeter swans persist in Yellowstone National Park? Paper presented at 25th Trumpeter Swan Society Conference, November 19–20, 2019, Alton, Illinois.

Shields, R. 1986. Management of the Tristate Swan subpopulation. *Trumpeter Swan Society Conference* 9: 5–6.

Siferd, T. D. 1982. Mink, Mustela vison, attacks trumpeter swan, *Cygnus buccinator*, cygnet. *Canadian Field-Naturalist* 96(3): 357–358.

Sladen, W. J. L., and D. L. Rininger. 2000. Teaching trumpeter swans pre-selected migration routes using ultralight aircraft as surrogate parents—second experiment, 1998-1999. *Trumpeter Swan Society Conference* 17: 63–165.

Sladen, W. J. L., and G. H. Olsen. 2007. Teaching geese, swans and cranes pre-selected migration routes using ultralight aircraft, 1990-2004—looking into the future. *Trumpeter Swan Society Conference* 20: 53–54.

Sladen, W. J. L., and J. C. Whissel. 2007. The winter distribution of trumpeter swans in relation to breeding areas: The first neckband study, 1972–81. *Trumpeter Swan Society Conference* 20: 117–125.

Sladen, W. J. L., and R. J. Limpert. 1992. A new look at the coded color neck and tarsus band protocol for North American swans. *Trumpeter Swan Society Conference* 13: 92–101.

Slater, G. L. 2006. *Trumpeter Swan* (Cygnus buccinator)*: A Technical Conservation Assessment.* USDA Forest Service, Rocky Mountain Region, Species Conservation Project. 39 pp. https://www.fs.usda. gov/Internet/FSE_DOCUMENTS/stelprdb5182067.pdf

Smith, C. S. 1996. Pacific Coast Joint Venture projects that secure or enhance swan habitat. *Trumpeter Swan Society Conference* 15: 133–135.

Smith, D. W., and N. Chambers. 2011. *The Future of Trumpeter Swans in Yellowstone National Park: Final Report Summarizing Expert Workshop, April 26–27, 2011.* Yellowstone National Park, WY: Yellowstone Center for Resources, National Park Service.

Smith, J. W. 1988. Status of Missouri's experimental trumpeter swan restoration program. *Trumpeter Swan Society Conference* 10: 100–103.

———. 1990. Trumpeter swan status report for Missouri. *Trumpeter Swan Society Conference* 11: 56–57.

Smith, J. W., and J. D. Wilson. 1986. Experimental restoration of trumpeter swans to Missouri. *Trumpeter Swan Society Conference* 9: 112–115.

Smith, M. C., J. M. Grassley, C. E. Grue, M. Davison, J. Bohannon, C. Schneider, and L. Wilson. 2007. Mortality of swans due to ingestion of lead shot, Whatcom County, Washington, and Sumas Prairie, British Columbia. *Trumpeter Swan Society Conference* 20: 114–116.

Smith, P. 2013. Status of the Trumpeter Swan (*Cygnus buccinator*) in Alberta: Update 2013. Alberta Wildlife Status Report No. 26. Edmonton: Alberta Environment and Sustainable Resource Development. 47 pp.

Snyder, J. W. 1990. The wintering and foraging ecology of the trumpeter swan, Harriman State Park, Idaho. *Trumpeter Swan Society Conference* 11: 6–8.

———. 1991a. Trumpeter swan winter habitat use on the Henry's Fork. *Trumpeter Swan Society Conference* 12: 174.

———. 1991b. The wintering and foraging ecology of the trumpeter swan, Harriman State Park of Idaho. MS thesis. Pocatello: Idaho State University.

Sojda, R. S., J. E. Cornely, and A. E. Howe. 2002. Development of an expert system for assessing trumpeter swan breeding habitat in the northern Rocky Mountains. Special Publication 1: Proceedings of the Fourth International Swan Symposium 2001 (E. C. Rees, S. L. Earnst, and J. Coulson, eds.) *Waterbirds* 25: 313–318.

Sojda, R. S., J. E. Cornely, L. H. Fredrickson, and A. E. Howe. 2000. Current research efforts in decision support system technology as applied to trumpeter swan management. *Trumpeter Swan Society Conference* 17: 139–144.

Spaeth, K. 2014. Reproductive success of trumpeter swans in west central Minnesota. MS thesis. Bemidji, MN: Bemidji State University. 41 pp.

Squires, J. R. 1991a. Trumpeter swan food habits, forage processing, activities, and habitat use. PhD diss. Laramie: University of Wyoming.

———. 1991b. The movements, productivity, and habitat-use patterns of trumpeter swans in the Greater Yellowstone area. *Trumpeter Swan Society Conference* 12: 172–173.

Squires, J. R., and S. H. Anderson. 1995. Trumpeter swan (*Cygnus buccinator*) food habits in the Greater Yellowstone ecosystem. *American Midland Naturalist* 133: 274–282.

———. 1997. Changes in trumpeter swan (*Cygnus buccinator*) activities from winter to spring in the Greater Yellowstone area. *American Midland Naturalist* 138: 208–214.

Squires, J. R., S. H. Anderson, and D. C. Lockman. 1992. Habitat selection of nesting and wintering trumpeter swans. Pp. 665–675 in *Wildlife 2001: Populations* (D. R. McCullough and R. H. Barrett, eds.). New York: Elsevier Applied Sciences.

Stearns, F. D., S. Breeser, and D. Sowards. 1990. Population expansion of trumpeter swans in the upper Tanana Valley, Alaska, 1982–1987. *Trumpeter Swan Society Conference* 11: 99–105.

St. Louis, M. J. 1994. Status of Oregon's trumpeter swan program. *Trumpeter Swan Society Conference* 14: 122–130.

Subcommittee on the Interior Population of Trumpeter Swans. 1997. Mississippi and Central Flyway Management Plan for the Interior Population of Trumpeter Swans. Unpublished report. Mississippi and Central Flyway Councils. Twin Cities, MN: US Fish and Wildlife Service.

Tangermann, H. L. 2002. Factors affecting the harvest vulnerability of trumpeter swans. MS thesis. Logan: Utah State University.

Tessman, S. A. 2000. Pacific Flyway Study Committee perspective on RMP [Rocky Mountain Population] trumpeter swan restoration. *Trumpeter Swan Society Conference* 17: 67–73.

Trumpeter Swan Society. 2010. Reporting marked trumpeter swans—collars, wing-tags, and bands. *Trumpeter Swan Society Blog*. https://trumpeterswansociety.wordpress.com/2010/01/26/reporting-marked-or-banded-trumpeter-swans/

Tori, G. 1997. Majestic trumpeters return to Ohio. *North American Swans* 26: 23–24.

Tori, G. M. 1999. Ohio's trumpeter swan restoration project—first year summary. *Trumpeter Swan Society Conference* 16: 14–17.

Trost, R. E. 1999. Trumpeter swan Rocky Mountain Population range expansion and tundra swan hunting: Is there a middle ground? *Trumpeter Swan Society Conference* 16: 59–60.

Trost, R. E., J. E. Cornely, and J. B. Bortner. 2000. US Fish and Wildlife Service Perspective on RMP [Rocky Mountain Population] trumpeter swan restoration. *Trumpeter Swan Society Conference* 17: 60–66.

Trumpeter Swan Society. n.d. Similar species and hybrids. *The Trumpeter Swan Society*. https://www.trumpeterswansociety.org/swan-information/identification/similar-species-and-hybrids.html. Accessed March 10, 2020. (See last section: Hybrids.)

Turner, B. 1988. Summary of results of Grande Prairie trumpeter swan collaring program (abstract). *Trumpeter Swan Society Conference* 10: 28.

Turner, B. C., and R. H. Mackay. 1982. The Population Dynamics of the Trumpeter Swans of Grand Prairie, Alberta. Unpublished report. Edmonton, AB: Canadian Wildlife Service.

US Fish and Wildlife Service. 2016. *Trumpeter Swan Survey of the Rocky Mountain Population, US Breeding Segment, Fall 2015*. Lakewood, CO: Migratory Birds and State Programs, Mountain-Prairie Region.

———. 2017. *Trumpeter Swan Survey of the Rocky Mountain Population, US Breeding Segment, Fall 2016*. Lakewood, CO: US Department of the Interior, Fish and Wildlife Service. 26 pp. and appendices.

US Fish and Wildlife Service et al. 1994. The North American Trumpeter Swan Status Report—1990. Unpublished cooperative report. US Fish and Wildlife Service, Canadian Wildlife Service, and the Trumpeter Swan Society.

Van Kirk, R., and R. Martin. 2000. Interactions among waterfowl herbivory, aquatic vegetation, fisheries, and flows below Island Park Dam, Idaho. *Trumpeter Swan Society Conference* 17: 85–94.

Varner, D. M. 2008. Survival and foraging ecology of Interior Population trumpeter swans. MS thesis. Carbondale: University of Illinois.

Varner, D. M., and M. W. Eichholz. 2012. Annual and seasonal survival of trumpeter swans in the upper Midwest. *Journal of Wildlife Management* 76: 129–135.

Vrtiska, M., and S. Comeau. 2009. *Trumpeter Swan Survey of the High Plains Flock, Interior Population, Winter 2008*. Lincoln: Nebraska Game and Parks Commission, and Martin, SD: US Fish and Wildlife Service, Lacreek National Wildlife Refuge. 14 pp. https://digitalcommons.unl.edu/usfwspubs/485/

Vrtiska, M. P., J. L. Hansen, and D. E. Sharp. 2007. Central Flyway perspectives on trumpeter swans. *Trumpeter Swan Society Conference* 20: 29–33.

Wareham, W., and G. Fowler. 1994. The Comox Valley waterfowl management project, 1991-93. *Trumpeter Swan Society Conference* 14: 93-94.

White, M. 2004. Habitat and management trends affecting trumpeter swans in Alberta. *Trumpeter Swan Society Conference* 19: 70-81.

White, M., and R. White. 2000. Rocky Mountain Population of trumpeter swans: Habitat trends in the Grande Prairie region. *Trumpeter Swan Society Conference* 17: 128-133.

Whitman, C. L., and C. D. Mitchell. 2004. Winter trumpeter swan mortality in southwestern Montana, eastern Idaho, and northwestern Wyoming, November 2000 through January 2003. *Trumpeter Swan Society Conference* 19: 96-100.

Will, G. C. 1991. Status of the Rocky Mountain Population of trumpeter swans, the range expansion project, and conditions in Idaho. *Trumpeter Swan Society Conference* 12: 177-180.

Wilson, L. K., J. E. Elliott, K. M. Langelier, A. M. Scheuhammer, and V. Bowes. 1998. Lead poisoning of trumpeter swans, *Cygnus buccinator*, in British Columbia, 1976-1994. *Canadian Field-Naturalist* 112(2): 204-211.

Wolfson, D. 2019. Interior Population trumpeter swan migration ecology and conservation. Paper presented at 25th Trumpeter Swan Society Conference, November 19-20, 2019, Alton, Illinois.

Whooper Swan

Abe, M. 1968. Some notes on the swans and on the main factors that caused their extensive death at Odaito Bay, Nemuro, Hokkaido. *Tori* 18, 379-91.

Albertsen, J. O., Y. Abe, S. Kashikawa, A. Ookawara, and K. Tamada. 2002. Age and sex differences in biometrics data recorded for whooper swans wintering in Japan. Special Publication 1: Proceedings of the Fourth International Swan Symposium 2001 (E. C. Rees, S. L. Earnst, and J. Coulson, eds.). *Waterbirds* 25: 334-339.

Black, J. M., and E C. Rees, 1984. The structure and age of the whooper swan population wintering at Caerlaverock, Dumfries, and Galloway, Scotland: An introductory study. *Wildfowl* 35: 21-36.

Boiko, D., H. Kampe-Persson, and J. Morknas. 2014. Breeding whooper swans *Cygnus cygnus* in the Baltic states, 1973-2013: Results of a recolonisation. *Wildfowl* 64: 207-216.

Boyd, H., and S. K. Eltringham. 1962. The whooper swan in England. *Bird Study* 9: 227-241.

Brazil, M. 2003. *The Whooper Swan*. London: T & AD Poyser. 512 pp.

Brazil, M. A. 1981. Geographical variation in the bill patterns of whooper swans. Wildfowl 32: 129-131.

————. 1981a. The behavioural ecology of *Cygnus cygnus cygnus* in central Scotland. Pp. 161-169 in *Proceedings of the Second International Swan Symposium, Sapporo, Japan, 1980* (G. V. T. Matthews and M. Smart, eds.). Slimbridge, UK: International Waterfowl Research Bureau.

————. 1981b. Summer behaviour of *Cygnus cygnus cygnus* in Iceland. Pp. 272-273 in *Proceedings of the Second International Swan Symposium, Sapporo, Japan, 1980* (G. V. T. Matthews and M. Smart, eds.). Slimbridge, UK: International Waterfowl Research Bureau.

————. 1981c. The behavioural ecology of the whooper swan (*Cygnus cygnus cygnus*). PhD diss. Stirling, Scotland: Stirling University.

————. 1981d. Geographical variation in the bill patterns of whooper swans. *Wildfowl* 32: 129-131.

————. 1984a. Winter feeding methods of the whooper swan (*Cygnus cygnus*). *Journal of the Yamashina Institute for Ornithology* 16: 83-86.

————. 1984b. The behaviour of whooper swans (*Cygnus cygnus*) wintering in a tidal environment. *Strix* 3: 40-49.

————. 1984c. The year of the whooper swan. *Birds* 10(4): 42-45.

————. 2002a. An addition to the diet of the whooper swan *Cygnus cygnus* from Japan. *Journal of the Yamashina Institute for Ornithology* 33: 210-212.

Brazil, M. A., and C. J. Spray. 1983. Large clutch and brood sizes of whooper swans. *Scottish Birds* 12: 226-227.

Brazil, M. A., and J. Shergalin. 2002a. The status and distribution of the whooper swan *Cygnus cygnus* in Russia I. *Journal of the Yamashina Institute for Ornithology* 34: 162–199.

———. 2002b. The status and distribution of the whooper swan *Cygnus cygnus* in Russia II. *Journal of the Yamashina Institute for Ornithology* 34: 279–308.

Byrd, G. V., D. L. Johnson, and D. D. Gibson. 1974. The birds of Adak Island, Alaska. *Condor* 76: 288–300.

Einarsson, Ó. 1996. Breeding biology of the whooper swan and factors affecting its breeding success, with notes on its social dynamics and life cycle in the wintering range. PhD diss. Bristol, UK: University of Bristol.

Einarsson, Ó., and E. C. Rees. 2002. Occupancy and turnover of whooper swans on territories in northern Iceland: Results of a long-term study. Special Publication 1: Proceedings of the Fourth International Swan Symposium 2001 (E. C. Rees, S. L. Earnst, and J. Coulson, eds.). *Waterbirds* 25: 202–210.

Fay, F. H. 1960. The distribution of waterfowl to St. Lawrence Island, Alaska. *Wildfowl Trust Annual Report* 12: 70–80.

Gardarsson, A., and K. H. Skarphedinsson. 1984. A census of the Icelandic whooper swan population. *Wildfowl* 35: 37–47.

Gibson, D. D., and B. Kessel. 1997. Inventory of the species and subspecies of Alaska birds. *Western Birds* 28: 45–95.

Gibson, D. D., and G. V. Byrd. 2007. *Birds of the Aleutian Islands, Alaska*. Cambridge, Massachusetts: Nuttall Ornithological Club, and Washington, DC: American Ornithologists' Union.

Haapenen, A. 1991.Whooper swan *Cygnus c. cygnus* population dynamics in Finland. *Wildfowl* (Supplement No. 1): 137–141.

Haapenen, A., M. Helminen, and H. K. Soumalainen. 1973a. The spring arrival and breeding phenology of the whooper swan *Cygnus c. cygnus* in Finland. *Finnish Game Research* 33: 3–38.

———. 1973b. Population growth and breeding biology of the whooper swan *Cygnus c. cygnus* in Finland in 1950-1970. *Finnish Game Research* 33: 39–60.

———. 1977. The summer behavior and habitat use of the whooper swan. *Finnish Game Research* 36: 49–81.

Hall, C., O. Crowe, G. M. C. Elwane, O. Einarsson, N. Calbrade, and E. Rees. 2016. Population size and breeding success of the Icelandic whooper swan *Cygnus cygnus*: Results of the 2015 International Census. *Wildfowl* 66: 75–97.

Hall-Craggs, J. 1974. Controlled antiphonal calling by whooper swans. *Ibis* 116: 218–32.

Hewson, R. 1964. Herd composition and dispersion in the whooper swan. *British Birds* 37: 26–31.

Höhn, E. O. 1948. Courtship display and species recognition in whooper swan. *British Birds* 41: 54.

Johnson, J. C., and W. J. L. Sladen. 1983. Whooper swans released in Maryland. *Maryland Birdlife* 39: 3–4.

Kakizawa, R. 1981. Hierarchy in the family group and social behaviour in wintering *Cygnus cygnus*. Pp. 210–11 in *Proceedings of the Second International Swan Symposium, Sapporo, 1980* (G. V. T. Matthews and M. Smart, eds.). Slimbridge, UK: International Waterfowl Research Bureau.

Kenyon, K. 1961. Birds of Amchitka Island, Alaska. *Auk* 78: 305–325.

———. 1963. Further observations of whooper swans in the Aleutian Islands. *Auk* 80: 540–542.

Knudsen, H. L., B. Laubeck, and A. Ohtonen. 2002. Growth and survival of whooper swan cygnets reared in different habitats in Finland. Special Publication 1: Proceedings of the Fourth International Swan Symposium 2001 (E. C. Rees, S. L. Earnst, and J. Coulson, eds.). *Waterbirds* 25: 211–220.

Laubek, B. 1998. The northwest European whooper swan (*Cygnus cygnus*) population: Ecological and management aspects of an increasing waterfowl population. PhD diss. Åarhus, Denmark: University of Åarhus.

Laubek, B., L. Nilsson, M. Wieloch, K. Koffijberg, C. Sudfelt, and A. Follestad. 1999. Distribution, number, and habitat choice of the Northwestern European whooper swan (*Cygnus cygnus*) population: Results of an international whooper swan census January 1995. *Vögelwelt* 120: 141–150.

McEneaney, T. 2004. A rare whooper swan at Yellowstone National Park, with comments on North American reports of the species. *North America Birds* 58: 301–308.

Murie, O. J. 1959. *Fauna of the Aleutian Islands and Alaska Peninsula.* North American Fauna No. 61. Washington, DC: US Department of the Interior, Fish and Wildlife Service. 406 pp.

Newth, J. L., E. C. Rees, R. L. Cromie, R. A. McDonald, S. Bearhop, D. J. Pain, G. L. Norton, C. Deacon and G. M. Hilton. 2016. Widespread exposure to lead affects the body condition of free-living whooper swans *Cygnus cygnus* wintering in Britain. *Environmental Pollution* 209: 60-67.

Nilsson, L. 2014. Long-term trends in the number of whooper swans *Cygnus cygnus* breeding and wintering in Sweden. *Wildfowl* 64: 197-206.

O'Donoghue, P. D., and J. O'Halloran. 1994. The behaviour of a wintering flock of whooper swans *Cygnus cygnus* at Rostellan Lake, Cork. *Proceedings of the Royal Irish Academy* 94: 109-118.

Rees, E. C., J. Bruce, and G. T. White. 2005. Factors affecting the behavioural responses of whooper swans (*Cygnus c. cygnus*) to various human activities. *Biological Conservation* 121: 369-382.

Rees, E. C., J. M. Black, C. J. Spray, and S. Thorisson. 1991. Comparative study of the breeding success of whooper swans, *Cygnus cygnus*, nesting in upland and lowland regions of Iceland. *Ibis* 133: 365-373.

Rees, E. C., J. S. Kirby, and A. Gilburn. 1997. Site selection by swans wintering in Britain and Ireland: The importance of geography, location, and habitat. *Ibis* 139: 337-352.

Salmonsen, E. 1950. *Grønlands Fugle/The Birds of Greenland.* Copenhagen: Munksgaard.

Stewart, A. G. 1978. Swans flying at 8,000 metres. *British Birds* 71: 459-460.

St. Louis, M. 1995. Whooper swan at Summer Lake Wildlife Area, Oregon, and California wintering areas. *Oregon Birds* 21: 35-37.

Sykes, P. W., Jr. and D. W. Sonneborn. 1998. First breeding records of whooper swan and brambling in North America at Attu Island, Alaska. *Condor* 100: 162-164.

Williamson, F. S. L., W. B. Emison, and C. M. White. 1971. *Studies of the Avifauna on Amchitka Island, Alaska, July 1, 1969-June 30, 1970.* Amchitka Bioenvironmental Program, Annual Progress Report. US AEC Report BMI-171-131. Battelle Memorial Institute, Columbus Laboratories.

Whistling (Tundra) Swan

Albertsen, J., R. Limpert, S. Earnst, W. Sladen, J. Hines, and T. Rothe. 1991. Demography of Eastern Population tundra swans *Cygnus columbianus columbianus.* Pp. 178-184 in *Wildfowl: Supplement Number 1, Third IWRB International Swan Symposium, Oxford, England, December 9-13, 1989* (J. Sears and P. J. Bacon, eds.). Slimbridge, UK: International Waterfowl and Wetlands Research Bureau.

Babcock, C. A., A. C. Fowler, and C. R. Ely. 2002. Nesting ecology of tundra swans on the coastal Yukon-Kuskokwim delta, Alaska. Special Publication 1: Proceedings of the Fourth International Swan Symposium (E. C. Rees, S. L. Earnst, and J. Coulson, eds.). *Waterbirds* 25): 236-240.

Badzinski, S. S. 2003. Dominance relations and agonistic behaviour of tundra swans (*Cygnus columbianus columbianus*) during fall and spring migration. *Canadian Journal of Zoology* 81: 727-733.

———. 2005. Social influences on tundra swan activities during migration. *Waterbirds* 28: 316-325.

Bart, J., and J. D. Nichols. 1992. Movements of tundra swans on the East Coast in winter (abstract). *Trumpeter Swan Society Conference* 13: 52.

Bart, J., and S. L. Earnst. 1991. Use of wetlands by grazing waterfowl in northern Alaska during late summer. *Journal of Wildlife Management* 55: 564-568.

Bart, J., R. Limpert, S. Earnst, W. Sladen, J. Hines, and T. Rothe. 1991. Demography of eastern population tundra swans *Cygnus columbianus columbianus.* Pp. 178-184 in *Wildfowl: Supplement Number 1, Third IWRB International Swan Symposium, Oxford, England, December 9-13, 1989* (J. Sears and P. J. Bacon, eds.). Slimbridge, UK: International Waterfowl and Wetlands Research Bureau.

Bartonek, J. C., J. R. Serie, and K. A. Converse. 1991. Mortality in tundra swans *Cygnus columbianus.* Pp. 356-358 in *Wildfowl: Supplement Number 1, Third IWRB International Swan Symposium, Oxford, England, December 9-13, 1989* (J. Sears and P. J. Bacon, eds.). Slimbridge, UK: International Waterfowl and Wetlands Research Bureau.

Bortner, J. B. 1985. Bioenergetics of wintering tundra swans in the Mattamuskeet region of North Carolina. MS thesis. College Park: University of Maryland.

———. 1988. Bioenergetics of wintering tundra swans in the Mattamuskeet region of North Carolina (abstract). *Trumpeter Swan Society Conference* 10: 158.

Boyd, S. 1994. Abundance patterns of trumpeter swans and tundra swans on the Fraser River delta, British Columbia (abstract). *Trumpeter Swan Society Conference* 14: 48.

Brackney, A. W., and R. J. King. 1993. *Aerial Breeding Pair Surveys of the Arctic Coastal Plain of Alaska: Revised Estimates of Waterbird Abundance, 1986–1992.* Unpublished report. Fairbanks, AK: US Fish and Wildlife Service, Migratory Bird Management. 21 pp.

Brandt, H. 1943. *Alaska Bird Trails.* Cleveland: Bird Research Foundation.

Canadian Wildlife Service Waterfowl Committee. 2017. *Population Status of Migratory Game Birds in Canada, November 2017.* CWS Migratory Birds Regulatory Report Number 49: Executive summaries. Ottawa, ON: Canadian Wildlife Service.

Caswell, D., J. Fuller, K. Gamble, J. Hansen, M. Huang, J. Johnson, T. Rothe, J. Serie, D. Sharp, and R. Trost. 2007. *Management Plan for the Eastern Population of Tundra Swans.* Report prepared for the Atlantic, Mississippi, Central, and Pacific Flyway Councils. 45 pp.

Chamberlain, E. B. 1966. Progress report: Productivity study of whistling swans wintering in Chesapeake Bay. Pp. 154–157 in *Proceedings of the 20th Annual Conference of the Southeastern Association of Game and Fish Commissioners.* http://www.seafwa.org/pdfs/articles/CHAMBERLAIN-154.pdf

Childress, D. 1991. Pacific Flyway Council comments on the draft position statement on tundra swan hunting. *Trumpeter Swan Society Conference* 12: 75.

Crawley, D. R., and E. G. Bolen. 2002. Effect of tundra swan grazing in winter wheat in North Carolina. Special Publication 1: Proceeding of the Fourth International Swan Symposium 2001 (E. C. Rees, S. L. Earnst, and J. Coulson, eds.). *Waterbirds* 25: 162–167.

Dau, C. P. 1981. Population structure and productivity of *Cygnus columbianus columbianus* on the Yukon, Alaska. Pp. 161–169 in *Proceedings of the Second International Swan Symposium, Sapporo, Japan, 1980* (G.V.T. Matthews and M. Smart, eds.). Slimbridge, UK: International Waterfowl Research Bureau.

———. 2014. Tundra swans of the lower Alaska Peninsula: A morphologically distinct North American population? (abstract). *Trumpeter Swan Society Conference* 23: (unpaged).

Dau, C. P., and J. E. Sarvis. 2002. Tundra swans of the lower Alaska Peninsula: Differences in migratory behavior and productivity. Special Publication 1: Proceedings of the Fourth International Swan Symposium 2001 (E. C. Rees, S. L. Earnst, and J. Coulson, eds.). *Waterbirds* 25: 241–249.

Dirksen, S., J. H. Beekman, and T. H. Slagboom. 1991. Bewick's swans *Cygnus columbianus bewickii* in the Netherlands: Numbers, distribution, and food choice during the wintering season. Pp. 228–237 in *Wildfowl: Supplement Number 1, Third IWRB International Swan Symposium, Oxford, England, December 9–13, 1989* (J. Sears and P. J. Bacon, eds.). Slimbridge, UK: International Waterfowl and Wetlands Research Bureau.

Earnst, S. L. 1992a. Behavior and ecology of tundra swans during summer, autumn, and winter. PhD diss. Columbus: Ohio State University.

———. 1992b. The timing of wing molt in tundra swans: Energetic and non-energetic constraints. *Condor* 94: 847–856.

———. 1992c. The habitat use of tundra swans (*Cygnus columbianus columbianus*) on an autumn migratory stopover (abstract). *Trumpeter Swan Society Conference* 13: 51.

———. 1994. Tundra swan habitat preferences during migration in North Dakota. *Journal of Wildlife Management* 58: 546–551.

———. 2002. Parental care in tundra swans during the pre-fledging period. Special Publication 1: Proceedings of the Fourth International Swan Symposium 2001 (E. C. Rees, S. L. Earnst, and J. Coulson, eds.). *Waterbirds* 25: 268–277.

Earnst, S. L., and J. Bart. 1991. Costs and benefits of extended parental care in tundra swans *Cygnus columbianus columbianus.* Pp. 260–267 in *Wildfowl: Supplement Number 1, Third IWRB*

International Swan Symposium, Oxford, England, December 9–13, 1989 (J. Sears and P. J. Bacon, eds.). Slimbridge, UK: International Waterfowl and Wetlands Research Bureau.

Earnst, S. L., and T. C. Rothe. 1994. Habitat preferences of tundra swans on their breeding grounds in northern Alaska (abstract). *Trumpeter Swan Society Conference* 14: 177.

Edwards, M. 1966. Wintering of whistling swans with mute swans in Grand Traverse Bay. *Jack-Pine Warbler* 14: 177.

Ely, C. R., D. A. Budeau, and U. G. Swain. 1987. Aggressive encounters between tundra swans and greater white-fronted geese during brood rearing. *Condor* 89: 420–422.

Ely, C. R., D. C. Douglas, A. C. Fowler, C. A. Babcock, D. V. Derksen, and J. Y. Takekawa. 1997. Migration behavior of tundra swans from the Yukon-Kuskokwim delta, Alaska. *Wilson Bulletin* 109: 679–692.

Ely, C. R., W. J. L. Sladen, H. M. Wilson, S. E. Savage, K. M. Sowl, B. Henry, M. Schwitters, and J. Snowdon. 2014. Delineation of tundra swan *Cygnus c. columbianus* populations in North America: Geographic boundaries and interchange. *Wildfowl* 64: 132–147.

Evans, M. E. 1977. Notes on the breeding behaviour of captive whistling swans by bill pattern. *Wildfowl* 28: 107–112.

Evans, M. E., and W. J. L. Sladen. 1980. A comparative analysis of the bill markings of whistling and Bewick's swans and out-of-range occurrences of the two taxa. *Auk* 97: 697–703.

Fischer, J. B., and R. A. Stehn. 2015. Nest population size and potential production of geese and spectacled eiders on the Yukon-Kuskokwim delta, Alaska, 2014. Unpublished report. Anchorage, AK: US Fish and Wildlife Service.

Gillette, L. N. 1992a. Position paper on tundra swan hunting: Introductory remarks. *Trumpeter Swan Society Conference* 13: 53–55.

———. 1992b. Potential techniques for monitoring the harvest in tundra swan hunts. *Trumpeter Swan Society Conference* 13: 81–84.

Hawkins, L. L. 1986a. Tundra swan (*Cygnus columbianus columbianus*) breeding behavior. MS thesis. St. Paul: University of Minnesota.

———. 1986b. Nesting behavior of male and female whistling swans and implications of male incubation. *Wildfowl* 37: 5–27.

Herbert, J. 1992. Summary of Montana's tundra swan hunting seasons, 1970–90. *Trumpeter Swan Society Conference* 13: 60–63.

Heyland, J. D., E. B. Chamberlain, C. F. Kimball, and D. H. Baldwin. 1970. Whistling swans breeding on the northwest coast of New Quebec. *Canadian Field-Naturalist* 84: 398–399.

Hines, J. E., and M. O. Wiebe Robertson, eds. 2001. *Surveys of Geese and Swans in the Inuvialuit Settlement Region, Western Canadian Arctic, 1989–2001.* Canadian Wildlife Service Occasional Paper No. 112. Ottawa, ON: Canadian Wildlife Service. 73 pp.

Hobson, K. A. 2004. Preliminary stable isotope analysis of tundra swan feathers: A new technique for delineating breeding origins of wintering birds. *Trumpeter Swan Society Conference* 19: 192–197.

Howie, R. R., and R. G. Bison. 2004. Wintering trumpeter and tundra swans in the southern interior of British Columbia. *Trumpeter Swan Society Conference* 19: 16–28.

Huener, J. D. 1992. Tundra swan hunting in Utah. *Trumpeter Swan Society Conference* 13: 76–80.

Johnson, M. A., and S. C. Kohn. 1991. Tundra swan hunting in North Dakota—results of the first season. *Trumpeter Swan Society Conference* 12: 65–72.

Jordan, M. 1991. Trumpeter and tundra swan survey in western Washington and Oregon—January 1989. *Trumpeter Swan Society Conference* 12: 14–17.

Kenow, K. P., J. M. Nissen, R. Drieslein, and E. M. Thorson. 2004. Tundra swan research needs on the upper Mississippi. *Trumpeter Swan Society Conference* 19: 180–189.

Klein, D. R. 1966. Waterfowl in the economy of the Eskimos on the Yukon-Kuskokwim Delta, Alaska. *Arctic* 19: 319–336.

Kohn, S. C., and M. A. Johnson. 1992. Results of tundra swan hunting seasons in North Dakota, 1988–90. *Trumpeter Swan Society Conference* 13: 64–73.

Lensink, C. J. 1973. Population structure and productivity of whistling swans on the Yukon delta, Alaska. *Wildfowl* 24: 21-25.

Limpert, R. J. 1974. Feeding preferences and behavior of whistling swans on the Upper Mississippi River. BS thesis. Winona, MN: St. Mary's College.

Limpert, R. J., and S. L. Earnst. 1994. Tundra swan (*Cygnus columbianus*). Version 2.0. *Birds of North America* (A. F. Poole and F. B. Gill, eds.). Ithaca, NY: Cornell Laboratory of Ornithology. https://birdsna.org/Species-Account/bna/species/tunswa/introduction

Limpert, R. J., H. A. Allen Jr., and W. J. L. Sladen. 1987. Weights and measurements of wintering tundra swans. *Wildfowl* 38: 108-113.

Limpert, R. J., W. J. L. Sladen, and H. A. Allen Jr. 1991. Winter distribution of tundra swans *Cygnus columbianus columbianus* breeding in Alaska and western Canadian Arctic. Pp. 78-83 in *Wildfowl: Supplement Number 1, Third IWRB International Swan Symposium, Oxford, England, December 9-13, 1989* (J. Sears and P. J. Bacon, eds.). Slimbridge, UK: International Waterfowl and Wetlands Research Bureau.

Lynch, J. J. 1972. Productivity and Mortality among Geese, Swans, and Brant: Part II. Historical Records from Productivity Appraisals, 1950-1971. Unpublished report. Laurel, MD: Bureau of Sport Fisheries and Wildlife, Patuxent Wildlife Research Center.

Mallek, E. J., R. Platte, and R. Stehn. 2006. Aerial Breeding Pair Surveys of the Arctic Coastal Plain of Alaska—2006. Unpublished report. Fairbanks, AK: US Fish and Wildlife Service, Migratory Bird Management.

McLaren, M. A., and P. L. McLaren. 1984. Tundra swans in northeastern Keewatin District, NWT. *Wilson Bulletin* 96: 6-11.

Mikami, 1989. First Japanese records of crosses between whistling *Cygnus columbianus columbianus* and Bewick's swans *Cygnus columbianus bewickii*. *Wildfowl* 40: 131-133.

Miller, S. L., M. A. Gregg, A. R. Kuritsubo, S. M. Combs, M. K. Murdock, J. A. Nilsson, B. R. Noon, and R. G. Botzler. 1988. Morphometric variation in tundra swans: Relationships among sex and age classes. *Condor* 90: 802-815.

Moermond, J. E., and M. A. Spindler. 1997. Migration route and wintering area of tundra swans *Cygnus columbianus* nesting in the Kobuk-Selawik lowlands, northwest Alaska. *Wildfowl* 48: 16-25.

Monda, M. J. 1991. Reproductive ecology of tundra swans on the Arctic Wildlife Refuge, Alaska. PhD diss. Moscow: University of Idaho.

Monda, M. J., J. T. Ratti, and T. R. McCabe. 1994a. Reproductive ecology of tundra swans on the Arctic National Wildlife Refuge, Alaska. *Journal of Wildlife Management* 58: 757-773.

———. 1994b. Behavioral responses of nesting tundra swans to human disturbance and implications for nest predation on the Arctic National Wildlife Refuge (abstract). *Trumpeter Swan Society Conference* 14: 178.

———. 1994c. Modification of tundra swan habitat by repeated use of nesting territories (abstract). *Trumpeter Swan Society Conference* 14: 179.

Munro, R. E. 1981a. Field feeding by *Cygnus columbianus columbianus* in Maryland. Pp. 261-272 in *Proceedings of the Second International Swan Symposium, Sapporo, Japan, 1980* (G. V. T. Matthews and M. Smart, eds.). Slimbridge, UK: International Waterfowl Research Bureau.

———. 1981b. Traditional return of *Cygnus columbianus columbianus* to wintering areas in Maryland's Chesapeake Bay. Pp. 81-98 in *Proceedings of the Second International Swan Symposium, Japan, 1980* (G. V. T. Matthews and M. Smart, eds.). Slimbridge, UK: International Waterfowl Research Bureau.

Nagel, J. 1965. Field feeding of whistling swans in northern Utah. *Condor* 67: 446-447.

Naves, L. C. 2015. *Alaska Subsistence Harvests of Birds and Eggs, 2013, Alaska Migratory Bird Co-Management Council.* Technical Paper No. 409. Alaska Department of Fish and Game, Division of Subsistence. http://www.adfg.alaska.gov/techpap/TP409.pdf

Nichols, J. D., J. Bart, R. J. Limpert, W. J. L. Sladen, and J. E. Hines. 1992. Annual survival rates of adult and immature eastern population tundra swans. *Journal of Wildlife Management* 56: 48-494.

Pacific Flyway Council. 1983. *Pacific Flyway Management Plan for the Western Population of Whistling Swans.* Pacific Flyway Study Committee. 27 pp.

———. 1989. *Pacific Flyway Management Plan for the Western Population of Whistling Swans.* Pacific Flyway Study Committee.

———. 2001. *Pacific Flyway Management Plan for the Western Population of Tundra Swans.* Portland, OR: Pacific Flyway Study Committee, US Fish and Wildlife Service. 28 pp. https://ecos.fws.gov/ServCat/DownloadFile/36959?Reference=36942

———. 2017. *Management Plan: Western Population of Tundra Swans.* Vancouver, WA: US Fish and Wildlife Service, Division of Migratory Bird Management. 27 pp.

Parmelee, D. F., and S. D. MacDonald. 1960. The birds of west-central Ellesmere Island and adjacent areas. *National Museums of Canada Bulletin* 169: 1–10l.

Parmelee, D. F., H. A. Stephens, and R. H. Schmidt. 1967. The birds of southeastern Victoria Island and adjacent small islands. *National Museums of Canada Bulletin* 222: 1–229.

Paulin, D. G. 1996. Tundra swan use in California's Central Valley. *Trumpeter Swan Society Conference* 15: 48–52.

Paulin, D. G., and E. Kridler. 1988. Spring and fall migration of tundra swans dyed at Malheur National Wildlife Refuge, Oregon. *Murrelet* 69: 1–9.

Petrie, S. A., and K. L. Wilcox. 2003. Migration chronology of Eastern Population tundra swans. *Canadian Journal of Zoology* 81: 861–870.

Petrie, S. A., S. S. Badzinski, and K. L. Wilcox. 2002. Population trends and habitat use of tundra swans staging at Long Point, Lake Erie. Special Publication 1: Proceedings of the Fourth International Swan Symposium 2001 (E. C. Rees, S. L. Earnst, and J. Coulson, eds.). *Waterbirds* 25: 143–149.

Rea, C., R. Ritchie, A. Stickney, and J. G. King. 2007. Multi-year monitoring program for tundra swans on the North Slope of Alaska. *Trumpeter Swan Society Conference* 20: 136–139.

Retterer, T. E. 1992. Nevada's tundra swan hunting program. *Trumpeter Swan Society Conference* 13: 56–59.

Ritchie, R. J., J. G. King, A. A. Stickney, B. A. Anderson, J. R. Rose, A. M. Wildman, and S. Hamilton. 2002. Population trends and productivity of tundra swans on the central Arctic Coastal Plain, northern Alaska, 1989–2000. Special Publication 1: Proceedings of the Fourth International Swan Symposium 2001 (E. C. Rees, S. L. Earnst, and J. Coulson, eds.). *Waterbirds* 25: 22–31.

Rozell, N. 2017. Tundra swans take two distinct paths to Alaska. *UAF News and Information*, University of Alaska Fairbanks, April 20, 2017. https://news.uaf.edu/tundra-swans-take-two-distinct-paths-to-alaska/

Serie, J. R., and J. B. Bartonek. 1991a. Harvest management of tundra swans *Cygnus columbianus columbianus* in North America. Pp. 359–367 in *Wildfowl: Supplement Number 1, Third IWRB International Swan Symposium, Oxford, England, December 9–13, 1989* (J. Sears and P. J. Bacon, eds.). Slimbridge, UK: International Waterfowl and Wetlands Research Bureau.

———. 1991b. Population status and productivity of tundra swans *Cygnus columbianus columbianus* in North America. Pp. 172–177 in *Wildfowl: Supplement Number 1, Third IWRB International Swan Symposium, Oxford, England, December 9–13, 1989* (J. Sears and P. J. Bacon, eds.). Slimbridge, UK: International Waterfowl and Wetlands Research Bureau.

Serie, J. R., D. Luszcz, and R. V. Raftovich. 2002. Population trends, productivity, and harvest of Eastern Population tundra swans. Special Publication 1: Proceedings of the Fourth International Swan Symposium 2001 (E. C. Rees, S. L. Earnst, and J. Coulson, eds.). *Waterbirds* 25: 32–36.

Sherwood, G. A. 1960. The whistling swan in the West with particular reference to Great Salt Lake Valley, Utah. *Condor* 62: 370–377.

Sladen, W. J. L. 1973. A continental study of whistling swans using neck collars. *Wildfowl* 24: 8–14.

———. 1991a. Swans should not be hunted. Pp. 348–375 in *Wildfowl: Supplement Number 1, Third IWRB International Swan Symposium, Oxford, England, December 9–13, 1989* (J. Sears and P. J. Bacon, eds.). Slimbridge, UK: International Waterfowl and Wetlands Research Bureau.

———. 1991b. Comments from a tundra swan researcher [on the draft position statement on tundra swan hunting]. *Trumpeter Swan Society Conference* 12: 76–77.

Sladen, W. J. L., and W. Cochran. 1969. Studies of the whistling swan, 1967–68. *Transactions of the North American Wildlife and Natural Resources Conference* 34: 42–50.

Spindler, M. A., and K. F. Hall. 1991. Local movements and habitat use of tundra or whistling swans *Cygnus columbianus* in the Kobuk-Selawik lowlands of northwest Alaska. *Wildfowl* 42: 17–32.

Stewart, D. B., and L. M. J. Bernier. 1989. Distribution, habitat, and productivity of tundra swans on Victoria Island, King William Island, and southwestern Boothia Peninsula, NWT. *Arctic* 42: 333–338.

Stewart, R. E. 1962. *Waterfowl Populations in the Upper Chesapeake Region.* Special Scientific Report—Wildlife 65. US Fish and Wildlife Service. 208 pp.

Stewart, R. E., and J. H. Manning. 1958. Distribution and ecology of whistling swans in the Chesapeake Bay region. *Auk* 75: 203–212.

Stickney, A. A., B. A. Anderson, R. J. Richie, and J. G. King. 2002. Spatial distribution, habitat characteristics and nest-site selection by tundra swans on the Central Arctic Coastal Plain, northern Alaska. Special Publication 1: Proceedings of the Fourth International Swan Symposium 2001 (E. C. Rees, S. L. Earnst, and J. Coulson, eds.). *Waterbirds* 25: 227–235.

Swystun, H. A., G. Kofinas, J. E. Hines, and R. D. Dawson. 2004. Local observations of tundra swans (*Cygnus columbianus columbianus*) in the Mackenzie delta region, Northwest Territories, Canada (abstract). *Trumpeter Swan Society Conference* 19: 190.

Swystun, H. A., R. D. Dawson, and J. E. Hines. 2004. Factors influencing reproductive success of tundra swans (*Cygnus columbianus columbianus*) (abstract). Trumpeter Swan Society Conference 19: 191.

Tate, D. J. 1966, Morphometric age and sex variation in the whistling swan (*Olor columbianus*). MS thesis. Lincoln: University of Nebraska-Lincoln.

Thompson, D. Q., and M. D. Lyons. 1964. Flock size in a spring concentration of whistling swans. *Wilson Bulletin* 76: 282–285.

Trost, R. E., D. Luszcz, T. C. Rothe, J. R. Serie, D. E. Sharp, and K. E. Gamble. 1999. Management and hunt plans for tundra swans. *Trumpeter Swan Society Conference* 16: 103–108.

Trumpeter Swan Society. 1991. The Trumpeter Swan Society position paper on tundra swan hunting—adopted January 1990. *Trumpeter Swan Society Conference* 12: 83–84.

Vaa, S. J. 1992. South Dakota tundra swan season—1990. *Trumpeter Swan Society Conference* 13: 74–75.

Vaa, S. J., M. A. Johnson, and J. L. Hansen. 1999. An evaluation of tundra swan hunting in the Central Flyway and concerns about trumpeter restoration. *Trumpeter Swan Society Conference* 16: 109–111.

Weaver, K. H. A. 2013. Tundra swan, *Cygnus columbianus columbianus*, habitat selection during the nonbreeding period. MS thesis. London: University of Western Ontario.

Wilk, R. J. 1987. Tundra swans in the Bristol Bay lowlands northern Alaska Peninsula. MS thesis. Stevens Point: University of Wisconsin-Stevens Point.

———. 1988. Distribution, abundance, population structure, and productivity of tundra swans in Bristol Bay, Alaska. *Arctic* 41: 288–292.

Wilkins, K. A., R. A. Malecki, P. J. Sullivan, J. C. Fuller, J. P. Dunn, L. J. Hindman, G. R. Costanzo, and D. Luszcz. 2010. Migration routes and bird conservation regions used by Eastern Population tundra swans (*Cygnus columbianus columbianus*) in North America. *Wildfowl* 60: 20–37.

Wilkins, K. A., R. A. Malecki, P. J. Sullivan, J. C. Fuller, J. P. Dunn, L. J. Hidman, G. R. Costanzo, S. A. Petrie, and D. Luszce. 2010. Population structure of tundra swans wintering in eastern North America. *Journal of Wildlife Management* 74: 1107–1111.

Wilkins, K., R. Malecki, S. Sheaffer, and D. Luszcz. 2001. Eastern Population tundra swans: Population status, survival, and movements. *North American Swans* 30: 15–18.

Wood, T., T. Brooks, and W. Sladen. 2002. Vocal characteristics of trumpeter and tundra swans and their hybrid offspring. Special Publication 1: Proceedings of the Fourth International Swan Symposium 2001 (E. C. Rees, S. L. Earnst, and J. Coulson, eds.). *Waterbirds* 25: 360–362.

Bewick's (Tundra) Swan

Bateson, P., W. Lotwick, and D. K. Scott. 1980. Similarity between the faces of parents and offspring in Bewick's swans and the differences between mates. *Journal of Zoology* (London) 191: 61–74.

Beekman, J., K. Koffijberg, J. Wahl, C. Hall, K. Devos, S. Pihl, B. Laubek, L. Luigujoe, M. Wieloch, H. Boland, S. Svarzas, L. Nilsson, A. Stipniece, V. Keller, P. Shimmings, and E. C. Rees. 2014. Long-term trends in the numbers and distribution of the Northwest European Bewick's swan population: Results of the international censuses (abstract). *Trumpeter Swan Society Conference* 23: (unpaged).

Beekman, J. H., M. R. Van Eerden, and S. Dirksen. 1991. Bewick's swans *Cygnus columbianus bewickii* utilising the changing resource of *Potamogeton pectinatus* during autumn in the Netherlands. Pp. 238-248 in *Wildfowl: Supplement Number 1, Third IWRB International Swan Symposium, Oxford, England, December 9-13, 1989* (J. Sears and P. J. Bacon, eds.). Slimbridge, UK: International Waterfowl and Wetlands Research Bureau.

Beekman, J. H., S. Dirkson, and T. H. Slagboom. 1985. Population size and breeding success of Bewick's swans wintering in Europe 1983–84. *Wildfowl* 36: 5–12.

Bowler, J. M. 1996. Feeding strategies of Bewick's swans (*Cygnus columbianus bewickii*) in winter. PhD diss. Bristol, UK: University of Bristol.

Bowler, J. M. 2005. Bewick's swan *Cygnus columbianus bewickii*. In *Bird Families of the World: Ducks, Geese, and Swans* (J. Kear, ed.). Oxford, UK: Oxford University Press.

Brazil, M. A. 2002. Brood amalgamation in Bewick's swan *Cygnus columbianus bewickii*: a record from Japan. *Journal of the Yamashina Institute of Ornithology* 33: 204–209.

Evans, M. E. 1975. Breeding behaviour of captive Bewick's swans. *Wildfowl* 26: 117–130.

———. 1976a. Aspects of the life cycle of the Bewick's swan, based on recognition of individuals at a wintering site. *Bird Study* 26: 149–162.

———. 1976b. Breeding behaviour of captive Bewick's swans. *Wildfowl* 26: 117–130.

———. 1977. Recognising individual Bewick's swans by bill pattern. *Wildfowl* 28: 153–158.

———. 1979a. Aspects of the life cycle of the Bewick's swan based on recognition of individuals at a wintering site. *Bird Study* 26: 149–162.

———. 1979b. The effect of weather on the wintering Bewick's swans *Cygnus columbianus bewickii* at Slimbridge, England. *Ornis Scandinavica* 10: 124–132.

———. 1980. The effects of experience and breeding status on the use of a wintering site by Bewick's swans *Cygnus columbianus bewickii*. *Ibis* 122: 287–297.

———. 1982. Movements of Bewick's swans *Cygnus columbianus bewickii* marked at Slimbridge, England from 1960 to 1979. *Ardea* 70: 59–75.

Evans, M. E., and J. Kear. 1978. Weights and measurements of Bewick's swans during winter. *Wildfowl* 29: 118–122.

Evans, M. E., and W. J. L. Sladen. 1980. A comparative analysis of the bill markings of whistling and Bewick's swans and out-of-range occurrences of the two taxa. *Auk* 97: 697–703.

Hayashi, T. 1982. Bill patterns of whistling swans *Cygnus columbianus jankowskii* wintering at Lake Suwa. *Tori: Bulletin of the Ornithological Society of Japan* 31: 1–16.

Johnstone, S. T. 1957. Breeding of Bewick's swans. *Avicultural Magazine* 63: 27–28.

Kondratiev, A. Ya. 1991. Breeding biology of Bewick's swans *Cygnus bewickii* in Chukotka, far eastern USSR. Pp. 167-171 in *Wildfowl: Supplement Number 1, Third IWRB International Swan Symposium, Oxford, England, December 9-13, 1989* (J. Sears and P. J. Bacon, eds.). Slimbridge, UK: International Waterfowl and Wetlands Research Bureau.

Mlodinow, S. G., and M. T. Schwitters. 2010. The status of Bewick's swan (*Cygnus columbianus bewickii*) in western North America. *North American Birds* 64(1): 2–13.

Rees, E. C. 1987. Conflict of choice within pairs of Bewick's swans regarding their migratory movement to and from the wintering grounds. *Animal Behaviour* 35: 1685–1693.

———. 1988. Aspects of the migration and movements of individual Bewick's swans. PhD diss. Bristol, UK: University of Bristol.

———. 1990. Bewick's swans: Their feeding ecology and coexistence with other grazing Anatidae. *Journal of Applied Ecology* 27: 939–951.

———. 2006. *Bewick's Swan.* London: T. & A. D. Poyser. 296 pp.

Rees, E. C., and P. J. Bacon. 1996. Migratory tradition in Bewick's swans *Cygnus columbianus bewickii.* Pp. 407–420 in *Proceedings Anatidae 2000 Conference, Strasbourg, France, 5–9 December 1994* (M. Birkan, J. van Vessem, P. Havet, J. Madsen, B. Trolliet, and M. Moser, eds.). *Gibier Faune Sauvage, Game Wildlife* 13.

Rees, E. C., J. S. Kirby, and A. Gilburn. 1997. Site selection by swans wintering in Britain. *Ibis* 139: 337–352.

Rose, P. M., and Scott, D. A.1994. *Waterfowl Population Estimates.* IWRB Publication No. 29. Slimbridge, UK.

Scott, D. K. 1967. The Bewick's swans at Slimbridge, 1966-67. *Wildfowl Trust Annual Report* 18: 24–27.

———. 1977. Breeding behaviour of wild whistling swans. *Wildfowl* 28: 101–106.

———. 1978a. Identification of individual Bewick's swans by bill patterns. Pp. 160–176 in *Animal Marking: Recognition Marking of Animals in Research* (B. Stonehouse, ed.). London: Macmillan.

———. 1978b. Social behaviour of wintering Bewick's swans. PhD diss. Cambridge, UK: University of Cambridge.

———. 1980a. Functional aspects of prolonged parental care in Bewick's swans. *Animal Behaviour* 28: 938–952.

———. 1980b. Functional aspects of the pair bond in winter in Bewick's swans (*Cygnus columbianus bewickii*). *Behavioural Ecology and Sociobiology* 7: 323–327.

———. 1980c. The behaviour of Bewick's swans (*Cygnus cygnus bewickii*) at the Welney Wildfowl Refuge, Norfolk, England, UK, and on the surrounding fens: A comparison. *Wildfowl* 31: 5–18.

———. 1980d. Winter behaviour of wild whistling swans: A comparison with Bewick's swans. *Wildfowl* 31: 119–121.

———. 1981. Geographical variation in the bill patterns of Bewick's swans. *Wildfowl* 32: 123–128.

———. 1988. Reproductive success in Bewick's swans. Pp. 220–236 in *Reproductive Success* (T. H. Clutton-Brock, ed.). Chicago: University of Chicago Press.

Scott, P. 1966. The Bewick's swans at Slimbridge. *Wildfowl Trust Annual Report* 17: 20–26.

Shchadilov, Y. M., A. V. Belousova, E. C. Rees, and J. M. Bowler. 1998. Long-term study of the nesting success in the Bewick's swans in the coastal tundra in the Nenetskiy Autonomous Okrug. *Casarca* 4: 217–228.

Solovyeva, D., and S. Vartanyan. 2014. Aspects of the breeding biology of Bewick's swans *Cygnus columbianus bewickii* nesting in high densities in the Chaun River delta, Chukotka, east Russia. *Wildfowl* 64: 148–166.

Syroechkovski, E. E. 2002. Distribution and population estimates for swans in the Siberian arctic in the 1990s. Special Publication 1: Proceedings of the Fourth International Swan Symposium 2001 (E. C. Rees, S. L. Earnst, and J. Coulson, eds.). *Waterbirds* 25: 100–113.

Syroechkovski, E. E., K. E. Litvin, and E. N. Guurtovaya. 2002. Nesting ecology of Bewick's swans on Vaygach Island, Russia. Special Publication 1: Proceedings of the Fourth International Swan Symposium 2001 (E. C. Rees, S. L. Earnst, and J. Coulson. eds.). *Waterbirds* 25: 221–226.

Wood, K. A., R. J. Nuijten, J. L. Newth, T. Haitjema, D. Vangeluwe, P. Ioannidis, A. L. Harrison, C. Mackenzie, G. M. Hilton, B. A. Nolet, and E. C. Rees, 2018. Apparent survival of an arctic-breeding migratory bird over 44 years of fluctuating population size. *Ibis* 160: 413–430.

Zea Books by Paul A. Johnsgard

The North American Swans: Their Biology and Behavior (2020)

The Abyssinian Art of Louis Agassiz Fuertes in the Field Museum (2020)

Wyoming's Ucross Ranch: Its Birds, History, and Natural Environment (2019)
by Jacqueline L. Canterbury and Paul A. Johnsgard

Wyoming Wildlife: A Natural History (2019)

A Naturalist's Guide to the Great Plains (2018)

The Birds of Nebraska, second edition (2018)

The Ecology of a Tallgrass Treasure: Audubon's Spring Creek Prairie (2018)

Common Birds of The Brinton Museum and Bighorn Mountains Foothills (2017)
by Jacqueline L. Canterbury and Paul A. Johnsgard

The North American Quails, Partridges, and Pheasants (2017)

The North American Perching and Dabbling Ducks: Their Biology and Behavior (2017)

The North American Whistling-Ducks, Pochards, and Stifftails (2017)

The North American Grouse: Biology and Behavior (2016)

The North American Geese: Their Biology and Behavior (2016)

The North American Sea Ducks: Their Biology and Behavior (2016)

Swans: Their Biology and Natural History (2016)

Birding Nebraska's Central Platte Valley and Rainwater Basin (2015)

At Home and at Large in the Great Plains: Essays and Memories (2015)

Global Warming and Population Responses among Great Plains Birds (2015)

Música de las Grullas: Una Historia Natural de las Grullas de América,
translated by Enrique H. Weir and Karine Gil-Weir (2014)

Birds and Birding in Wyoming's Bighorn Mountains Region (2013)
by Jacqueline L. Canterbury, Paul A. Johnsgard, and Helen F. Downing

Birds of the Central Platte River Valley and Adjacent Counties (2013)
by Mary Bomberger Brown and Paul A. Johnsgard

The Birds of Nebraska, revised edition (2013)

A Prairie's Not Scary (2012)

Wings over the Great Plains: Bird Migrations in the Central Flyway (2012)

Wetland Birds of the Central Plains: South Dakota, Nebraska, and Kansas (2012)

Rocky Mountain Birds: Birds and Birding in the Central and Northern Rockies (2011)

A Nebraska Bird-Finding Guide (2011)